ENDANGERED
ANIMALS
GRAPHIC
ARCHIVES

150

**멸종위기동물
그래픽
아카이브**

**사라져가는
동물들에 대한
기록**

시오시르 GRAPHICS
DESIGN BOUTIQUE

Index
목차

08 Preface 서문	36 Harp Seal 하프 물범	64 Mountain Tapir 산악맥
10 Northern Saw-whet Owl 애기금눈 올빼미	38 Golden Lion Tamarin 황금사자 타마린	66 Guanaco 구아나코
12 Emperor Tamarin 황제 타마린	40 Quotes about animals 동물에 대한 명언들	68 Jaguarundi 재규어 런디
14 Two-toed Sloth 두발가락나무늘보	42 Polar Bear 북극곰	70 Weddell Seal 웨델 바다표범
16 Common Marmoset 커먼 마모셋	44 Reindeer 순록	72 Quotes about animals 동물에 대한 명언들
18 Pygmy Raccoon 피그미라쿤	46 American Badger 아메리카 오소리	74 Gray-handed night monkey 회색손 올빼미 원숭이
20 Emperor Penguin 황제펭귄	48 Common Woolly Monkey 갈색양털 원숭이	76 Bald uakari 대머리 우아카리
22 Black-footed Ferret 검은발 페렛	50 Red Fox 붉은여우	78 Eastern Spotted Skunk 동부 얼룩 스컹크
24 Quotes about animals 동물에 대한 명언들	52 White-nosed Coati 흰코 코아티	80 Spectacled Bear 안경곰
26 Cotton-headed Tamarin 목화머리 타마린	54 Gray Wolf 회색늑대	82 Silky Anteater 애기 개미핥기
28 Margay 마게이	56 Quotes about animals 동물에 대한 명언들	84 Andean Cat 안데스고양이
30 Northern Rockhopper Penguin 북부바위뛰기 펭귄	58 Capybara 카피바라	86 Ring Tailed Cat 알락꼬리 고양이
32 Three-toed Sloth 세발가락나무늘보	60 Arctic Fox 북극여우	88 Quotes about animals 동물에 대한 명언들
34 Vancouver Island Marmot 밴쿠버섬마못	62 Blonde Capuchin 금발 카푸친	90 Island fox 아일랜드 여우

92 Bighorn Sheep 큰뿔양	126 Thomson's Gazelle 톰슨가젤	158 Indri 인드리 원숭이
94 Tehuantepec Jackrabbit 테우안테펙 산토끼	128 Red-bellied Lemur 붉은배 여우원숭이	160 African Golden Cat 아프리카 황금고양이
96 Galapagos fur seal 갈라파고스 물개	130 Fennec Fox 사막여우	162 Red-tailed Monkey 붉은꼬리 원숭이
98 Southern Tamandua 작은 개미핥기	132 Animal protection laws 국가별 동물보호법	164 Animal protection laws 국가별 동물보호법
100 Maned Wolf 갈기늑대	134 L'Hoest's Monkey 로에스트 원숭이	166 Crowned Lemur 관여우 원숭이
102 Blue-throated Macaw 푸른목 금강앵무	136 Lion 사자	168 Addax 아닥스
104 Animal protection laws 국가별 동물보호법	138 Meerkat 미어캣	170 Ethiopian Wolf 에티오피아늑대
106 Western Pygmy Marmoset 서부 피그미 마모셋	140 Ibex 아이벡스	172 Gelada 겔라다 개코원숭이
108 Caqueta Titi Monkey 카케타 티티	142 Gorilla 고릴라	174 Wolf's Mona Monkey 늑대모나 원숭이
110 American Marten 아메리카 담비	144 Otter 수달	176 African Wild Dog 아프리카 들개
112 Barn owl 원숭이 올빼미	146 Coquerel's Sifaka 코쿠렐 시파카	178 Ring-tailed Lemur 알락꼬리여우원숭이
114 Sea Otter 해달	148 Animal protection laws 국가별 동물보호법	180 Animal protection laws 국가별 동물보호법
116 Squirrel Monkey 다람쥐 원숭이	150 De Brazza's Monkey 브라자원숭이	182 Fossa 포사
120 Mandrill 맨드릴원숭이	152 Sand Cat 모래고양이	184 Red-fronted Brown Lemur 붉은이마 여우원숭이
122 Bateleur 달마수리	154 Iberian Lynx 이베리아 스라소니	186 Great Blue Turaco 큰 파랑부채머리새
124 Cheetah 치타	156 Aye-aye 아이아이 원숭이	188 Giant Dragon Lizard 자이언트 드래곤 도마뱀

Index
목차

190 Egyptian vulture
     이집트독수리

192 European ground squirrel
     유럽 땅다람쥐

194 Gerenuk
     게레눅

196 We've been saved!
     위기로부터 벗어난 동물들

198 Colobus
     콜로부스

200 Diana Monkey
     다이아나 원숭이

202 Western red colobus
     서부 붉은 콜로부스

204 Honduran white bat
     온두라스 흰박쥐

206 Moustached Monkey
     콧수염 원숭이

208 Bale Monkey
     베일 원숭이

210 Bongo
     봉고

212 We've been saved!
     위기로부터 벗어난 동물들

214 Caracal
     카라칼

216 Manatee
     매너티

218 Vervet Monkey
     버빗 원숭이

220 Striped Hyaena
     줄무늬 하이에나

222 Eurasian Eagle-owl
     수리부엉이

224 Atlantic Puffin
     대서양 푸핀

226 We've been saved!
     위기로부터 벗어난 동물들

228 Bearded Vulture
     수염수리

230 Marbled Polecat
     마블드 긴털 족제비

232 Malbrouck Monkey
     말브룩 원숭이

236 Koala
     코알라

238 Siberian Flying Squirrel
     하늘다람쥐

240 Platypus
     오리너구리

242 Bridled Nail-tail Wallaby
     고삐 발톱꼬리 왈라비

244 Common ringtail possum
     알락꼬리 주머니쥐

246 Greater Glider
     주머니 날다람쥐

248 We've been saved!
     위기로부터 벗어난 동물들

250 Bilby
     빌비

252 Quokka
     쿼카

254 Black Crested Gibbon
     검은손기번 원숭이

256 Snow Leopard
     설표

258 Slow Loris
     슬로우 로리스

260 Siberian Weasel
     족제비

262 Siberian Tiger
     시베리아 호랑이

264 We've been saved!
     위기로부터 벗어난 동물들

266 Proboscis Monkey
     코주부 원숭이

268 Sugar Glider
     슈가 글라이더

270 Japanese Marten
     산달

272 Japanese macaque
     일본 원숭이

274 Giant Panda
     자이언트팬더

276 Tawny Owl
올빼미

278 Siau Island Tarsier
시아우섬 안경원숭이

280 We've been saved!
위기로부터 벗어난 동물들

282 Lesser Panda
레서판다

284 Asiatic Black Bear
반달가슴곰

286 Tonkin Snub-nosed Monkey
통킹 들창코원숭이

288 Japanese Serow
일본 산양

290 Matschie's Tree Kangaroo
매치스 나무캥거루

292 Sun Bear
썬베어

294 Malay Civet
말레이 시벳

296 We've been saved!
위기로부터 벗어난 동물들

298 Indian Giant Squirrel
인도 큰다람쥐

300 Gaur
인도 들소

302 Black Javan Leopard
검은 자바 표범

304 Natuna Island Surili
나투나섬 잎원숭이

306 Sumatran Orangutan
수마트라 오랑우탄

308 Indian Muntjac
인도 문착

310 Ili Pika
일리 우는 토끼

312 We're being mistreated
우리는 억울하다

314 Javan Hawk-eagle
자바 뿔매

316 Water deer
고라니

318 Plateau Pika
고원 우는 토끼

320 Przewalski's Horse
프르제발스키 말

322 Binturong
빈투롱

324 Yellow-eyed Penguin
노란눈 펭귄

326 We're being mistreated
우리는 억울하다

328 Sulawesi Serpent-eagle
술라웨시 뱀독수리

330 Manul
마눌

332 Komodo Dragon
코모도 왕도마뱀

334 Flat-Headed Cat
납작머리 살쾡이

336 Golden Snub-nosed Monkey
황금 들창코 원숭이

338 Prevost's Squirrel
삼색 다람쥐

340 We're being mistreated
우리는 억울하다

342 Philippine Eagle
필리핀 독수리

344 Common Spotted Cuscus
얼룩 쿠스쿠스

346 Eurasian Red Squirrel
북방 청서

348 Eurasian Brown Bear
유라시아 불곰

350 Eurasian Badger
오소리

352 Dusky Pademelon
검은 덤불 왈라비

356 How to live together
더불어 살아가는 방법

360 Reference
참고 자료

Preface
서문

"디자인을 통해 세상을 좀 더 나은 곳으로 만들 수 있을까?" 멸종위기동물 그래픽아카이브는 디자인 회사의 작은 질문에서 시작된 프로젝트입니다. 사람이 생을 마감할때 자신의 영정 사진을 남기듯, 사라져가는 동물들의 초상화와 정보를 디자인으로 기록하는 이 프로젝트는 다양한 제품, 전시회, 기업과의 협업 캠페인으로 이어지고 있습니다. 이 책은 이전에 출간된 멸종위기동물 그래픽아카이브 01, 02를 모아 개정하고 새로운 내용을 증보한 것으로 150종의 멸종위기동물 그래픽과 생태 정보, 동물과 환경에 대한 여러 이야기를 담고 있습니다.

멸종 위기의 동물이 시사하는 바는 자연환경의 보전이라는 현 세대의 인류에게 큰 과제이자 안타까움만으로 방치할 수 없는 눈앞의 현실입니다. 이러한 디자인 프로젝트가 위기의 동물들을 직접 구호한다거나 종 보전을 위한 과학적 도구가 될 수는 없을 것입니다. 하지만 이러한 활동을 통해 좀 더 많은 사람들이 안타까움과 관심을 갖게 되고, 관심이 모여 목소리가 되고, 목소리가 모여 행동으로 이어지는 시발점이 되기를 희망합니다. 환경 운동가나 동물 보호 단체, 정부만 할 수 있는 일이 아닙니다. 당장 오늘부터 일회용품을 덜 써 보는 것만으로도 충분한 시작입니다. 약간의 관심과 각자의 자리에서 할 수 있는 작은 일들이 세상을 좀 더 나은 곳으로 바꿀 수 있을 것입니다.

이 책이 세상에 나올 수 있도록 도움을 주신 모든 분들께 감사의 마음을 전합니다.

2023년 봄
성실그래픽스 김남성

"Can we make the world a better place through design?" Endangered Animals Graphic Archives project started with a small question from a design company. Like leaving a portrait behind before passing away, this project records portraits and information about disappearing animals through design. The project has led to various products, exhibitions, and collaboration campaigns with companies. This book combines and revises the previous publications 01 and 02, & includes new content featuring 150 graphics & ecological information on endangered species, as well as various stories about animals and the environment.

What endangered species tell us is that protecting the environment is a task for this generation and cannot be left aside, just feeling bad for it. This design projects can not save the animals directly. However, we hope that such activities can be a stepping stone to bring in more attention to these issues, which will lead to a voice and to an act. It is not something that only environmental activists or certain groups can do. Using less disposable products from now on can also be a start. A bit of attention and small differences that we can make can change the world.

I'd like to thank everyone who helped make this book possible for publication.

Spring 2023
Sungsil Graphics, Namsung Kim

ENDANGERED
ANIMALS
GRAPHIC
ARCHIVES
150

- **EX** Extinct 절멸종
- **EW** Extinct in the Wild 야생절멸종
- **CR** Critically Endangered 위급종
- **EN** Endangered 위기종
- **VU** Vulnerable 취약종
- **NT** Near Threatened 준위협종
- **LC** Least Concern 관심대상종

레드리스트는 세계자연보전연맹 IUCN에서 지정하는 멸종위기의 동/식물들에 대한 분류학적 정보의 모음입니다.
이 책에 등장하는 동물들의 멸종위기 등급은 2023년을 기준으로 작성되었으며 이후 변경될 수 있습니다.
RedList is a collection of taxonomic information on endangered animals and plants designated by the IUCN
(International Union for Conservation of Nature and Natural resources) for conservation purposes.
The RedList levels of the animals featured in this book were written based on the year 2023 and
may be subject to change thereafter.

Northern saw-whet owl, unique shriek

신기한 울음소리, 애기금눈 올빼미

Size 크기	18~22cm / 0.065~0.1kg	**Lifespan** 평균수명	7~16 years
Diet 식성	Carnivore 육식	**Habitat** 서식지	Forest 숲
Behavior 행동	Solitary living 단독생활	**Distribution** 분포지역	North America 북아메리카
Reproduction 번식	Average 4~7eggs 평균 4~7개의 알	**Conservation status** 멸종위기 등급	Least concern (LC) 관심대상종

Northern saw-whet owl's English name was given because its screeching sounds were like see-sawing.
애기금눈 올빼미의 영어명 'saw-whet'은 울음소리가 마치 톱소리와 흡사하다고 하여 붙여졌습니다.

NORTHERN SAW-WHET OWL
LEAST CONCERN SPECIES

EX Extinct
EW Extinct in the Wild
CR Critically Endangered
EN Endangered
VU Vulnerable
NT Near Threatened
LC Least Concern

Emperor tamarin,
bearded emperor
황제의 수염, 황제 타마린

Size 크기	18~30cm / 0.2~0.9kg
Diet 식성	Omnivore 잡식
Behavior 행동	Group living 무리생활
Reproduction 번식	Average 2 평균 2마리
Lifespan 평균수명	8~15 years
Habitat 서식지	Rainforest 열대우림
Distribution 분포지역	Brazil, Peru, Bolivia 브라질, 페루, 볼리비아
Conservation status 멸종위기등급	Least concern (LC) 관심대상종

EMPEROR TAMARIN
LEAST CONCERN SPECIES

EX Extinct
EW Extinct in the Wild
CR Critically Endangered
EN Endangered
VU Vulnerable
NT Near Threatened
LC Least Concern

Two toed sloth,
slow animal in the world
세계에서 가장 느린 동물, 두발가락나무늘보

Size
크기

50~74cm / 4~8kg

Diet
식성

Herbivore
초식

Behavior
행동

Solitary living
단독 생활

Reproduction
번식

Average 1
평균 1마리

Lifespan
평균수명

28 years in captivity
사육시 평균 28년

Habitat
서식지

Tropical forest
열대 우림

Distribution
분포지역

Central America and Northern South America
중앙 아메리카 및 남아메리카 북부

Conservation status
멸종위기 등급

Least concern (LC)
관심대상종

TWO TOED SLOTH
LEAST CONCERN SPECIES

EX Extinct
EW Extinct in the Wild
CR Critically Endangered
EN Endangered
VU Vulnerable
NT Near Threatened
LC Least Concern

Common marmoset,
moving fan ear
움직이는 부채 귀, 커먼마모셋

Size 크기	14.5~20cm / 0.3~0.36kg
Diet 식성	Tree sap, insect, fruit 나무 수액, 곤충, 과일
Behavior 행동	Group living 무리 생활
Reproduction 번식	Average 2 평균 2마리
Lifespan 평균수명	10 years in captivity 사육시 평균 10년
Habitat 서식지	Rainforest 열대우림
Distribution 분포지역	Southeastern Brazilian coastal 브라질 남동부 연안
Conservation status 멸종위기 등급	Least concern (LC) 관심대상종

COMMON MARMOSET
LEAST CONCERN SPECIES

EX Extinct
EW Extinct in the Wild
CR Critically Endangered
EN Endangered
VU Vulnerable
NT Near Threatened
LC Least Concern

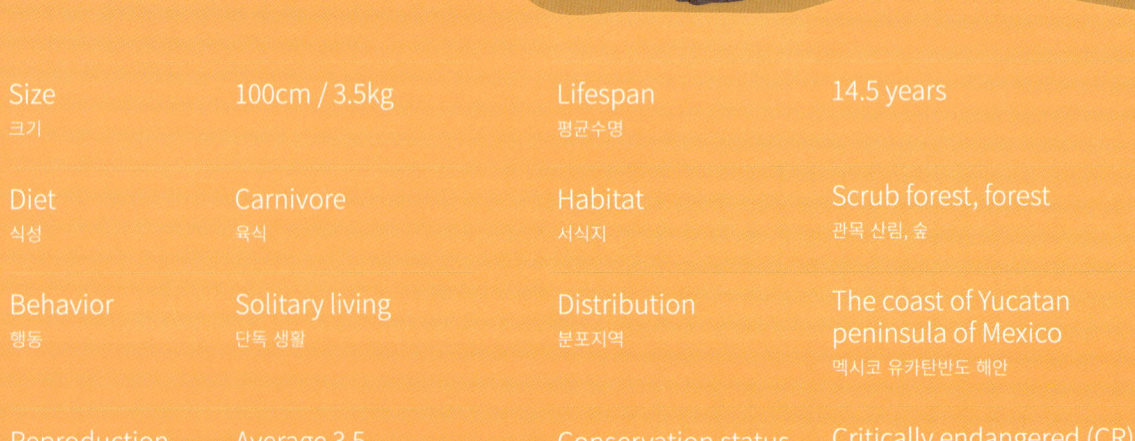

Pygmy raccoon,
the invader of trashcan

휴지통의 침입자, 피그미 라쿤

Size 크기	100cm / 3.5kg	Lifespan 평균수명	14.5 years
Diet 식성	Carnivore 육식	Habitat 서식지	Scrub forest, forest 관목 산림, 숲
Behavior 행동	Solitary living 단독 생활	Distribution 분포지역	The coast of Yucatan peninsula of Mexico 멕시코 유카탄반도 해안
Reproduction 번식	Average 3.5 평균 3.5마리	Conservation status 멸종위기 등급	Critically endangered (CR) 위급종

A pygmy raccoon mess up trash cans in North America.
북아메리카에서 거주자들의 쓰레기통을 엉망으로 만듭니다.

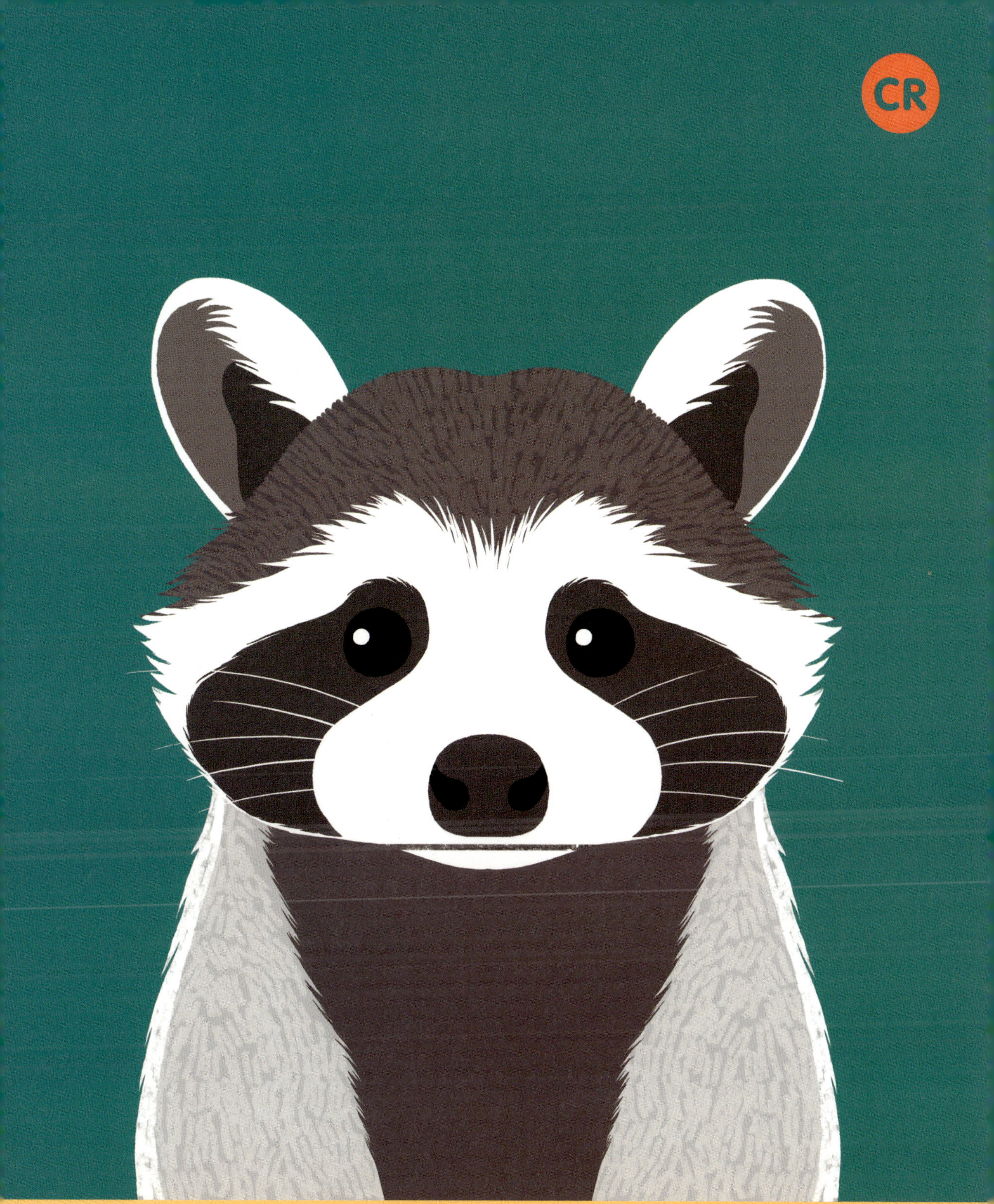

- EX Extinct
- EW Extinct in the Wild
- CR Critically Endangered
- EN Endangered
- VU Vulnerable
- NT Near Threatened
- LC Least Concern

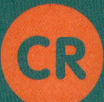

PYGMY RACCOON
CRITICALLY ENDANGERED SPECIES

Emperor penguin,
gentleman of the Antarctic

남극의 신사, 황제펭귄

Size 크기	115cm / 22~37kg
Diet 식성	Crustaceans, Fish 갑각류, 어류
Behavior 행동	Group living 무리 생활
Reproduction 번식	Average 1 평균 1마리
Lifespan 평균수명	20 years
Habitat 서식지	Ice cliff, iceberg 얼음 절벽, 빙산
Distribution 분포지역	Antarctica 남극
Conservation status 멸종위기 등급	Near threatened (NT) 준위협종

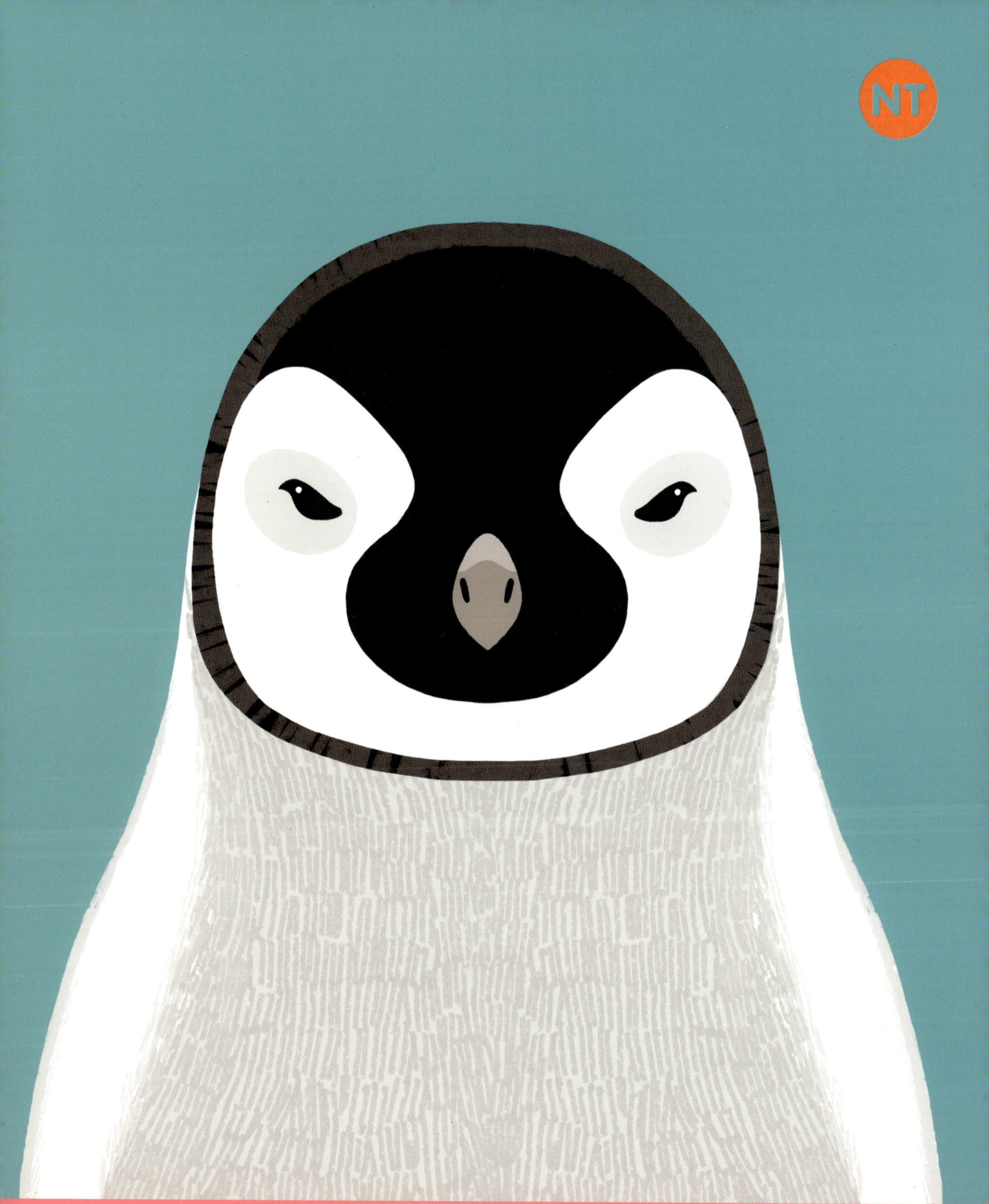

EMPEROR PENGUIN
NEAR THREATENED SPECIES

- EX Extinct
- EW Extinct in the Wild
- CR Critically Endangered
- EN Endangered
- VU Vulnerable
- NT Near Threatened
- LC Least Concern

Black footed ferret, the traveler of wildlife
야생의 여행가, 검은발 페렛

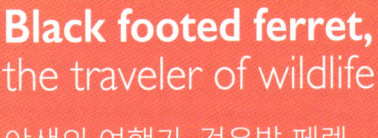

Size 크기	38~60cm / 0.65~1.4kg	Lifespan in captivity 사육 평균수명	12 years
Diet 식성	Small mammals 작은 포유류	Habitat 서식지	Savana, grassland 사바나, 초원
Behavior 행동	Solitary living 단독 생활	Distribution 분포지역	North America, Mexico 북아메리카, 멕시코
Reproduction 번식	Average 3 평균 3마리	Conservation status 멸종위기 등급	Endangered (EN) 위기종

A black-footed ferret travels long distances at a time.
이동할 때 한 번에 먼 거리를 이동합니다.

QUOTES ABOUT ANIMALS

The love for all living creatures is the most noble attribute of man.

살아있는 모든 만물을 사랑하는 것이야말로 인간의 가장 고귀한 특성이다.

Charles Robert Darwin, 1809~1882
an English biologist

동물에 대한 명사들의 한마디

We must fight against the spirit of unconscious cruelty with which we treat the animals.

우리는 동물들에게 무의식적으로 가하고 있는 잔혹성과 싸워야 합니다.

Albert Schweitzer, 1875~1965
a French-German organist, philosopher

Cotton headed tamarin, the piano maestro
피아노 거장, 목화머리 타마린

Size 크기	18~30cm / 0.2~0.9kg	Lifespan 평균수명	8~15 years
Diet 식성	Omnivore 잡식	Habitat 서식지	Tropical forest 열대우림
Behavior 행동	Group living 무리생활	Distribution 분포지역	Northwestern Colombia 콜롬비아 북서부 지역
Reproduction 번식	Average 2 평균 2마리	Conservation status 멸종위기등급	Critically endangered(CR) 위급종

Cotton-headed tamarin is called 'the Liszt monkey' because it has a similar hair style with the famous Hungarian composer Franz Liszt.
목화머리 타마린은 헝가리 피아노 거장 인 프란츠 리스트의 머리스타일과 비슷하다 하여 '리스트 원숭이'라고도 불립니다.

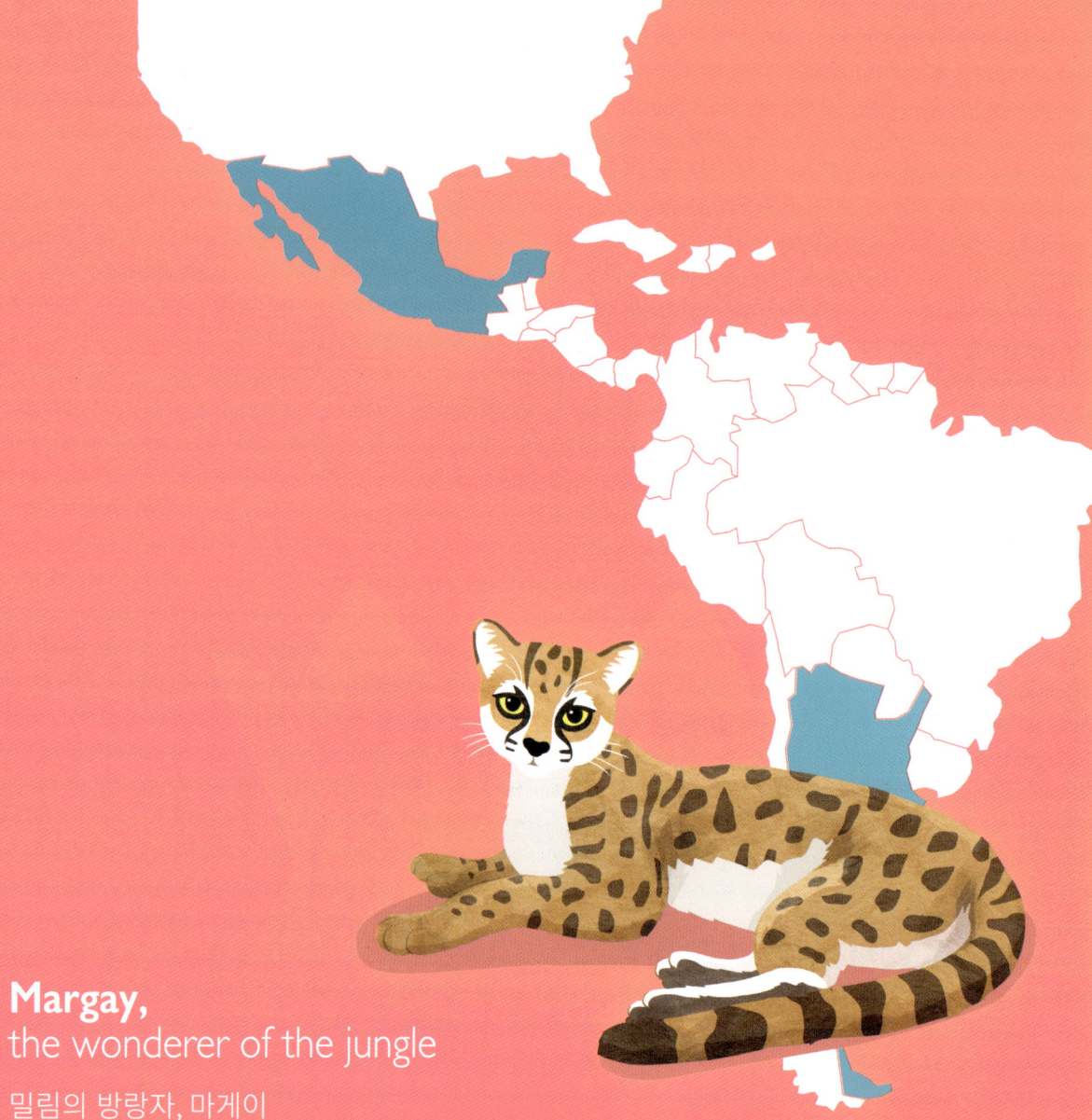

Margay,
the wonderer of the jungle
밀림의 방랑자, 마게이

Size 크기	46~79cm / 2~4kg	**Lifespan in captivity** 사육 평균수명	20 years
Diet 식성	Carnivore 육식	**Habitat** 서식지	Tropical forest 열대우림
Behavior 행동	Solitary living 단독 생활	**Distribution** 분포지역	Northern Mexico & Argentina 멕시코 북부 및 아르헨티나 북부
Reproduction 번식	Average 1.5 평균 1.5마리	**Conservation status** 멸종위기 등급	Near threatened (NT) 준위협종

A margay wanders Mexico to Argentina aimlessly.
멕시코에서 아르헨티나에 이르는 우림지대를 정처 없이 떠돌아다니면서 지냅니다.

MARGAY
NEAR THREATENED SPECIES

EX Extinct
EW Extinct in the Wild
CR Critically Endangered
EN Endangered
VU Vulnerable
NT Near Threatened
LC Least Concern

Northern rockhopper penguin,
penguin quicksilver

가장 민첩한 펭귄, 북부바위뛰기펭귄

Size 크기	45~58cm / 2~5kg		**Lifespan** 평균수명	10~15 years
Diet 식성	Krill, fish, squid 크릴새우, 생선, 오징어 등		**Habitat** 서식지	Rock cliff near the Ocean 바다 근처 암벽
Behavior 행동	Group living 무리생활		**Distribution** 분포지역	Tristan da Cunha, Gough island in the south Atlantic Ocean 대서양 남쪽 트리스탄다쿠나제도, 고프섬
Reproduction 번식	Average 2eggs 평균 2개의 알		**Conservation status** 멸종위기 등급	Endangered (EN) 위기종

Northern rockhopper penguin lives on a steep cliff and moves by jumping from a rock to rock.
북부바위뛰기펭귄은 가파른 절벽에 서식하며, 바위에서 바위로 점프하며 이동합니다.

NORTHERN ROCKHOPPER PENGUIN
ENDANGERED SPECIES

- EX Extinct
- EW Extinct in the Wild
- CR Critically Endangered
- EN Endangered
- VU Vulnerable
- NT Near Threatened
- LC Least Concern

Three-toed sloth,
champion of monkey bars
매달리기 챔피언, 세발가락 나무늘보

Size 크기	48~53cm / 2~4kg	**Lifespan** 평균수명	30~40 years
Diet 식성	Herbivore 초식	**Habitat** 서식지	Rainforest 열대우림
Behavior 행동	Solitary living 단독생활	**Distribution** 분포지역	South America & Central America 남아메리카와 중앙아메리카
Reproduction 번식	Average 1 평균 1마리	**Conservation status** 멸종위기 등급	Critically endangered (CR) 위급종

THREE-TOED SLOTH
CRITICALLY ENDANGERED SPECIES

EX Extinct
EW Extinct in the Wild
CR Critically Endangered
EN Endangered
VU Vulnerable
NT Near Threatened
LC Least Concern

Vancouver island marnot,
Olympics "sidekick" mascot

올림픽 보조 마스코트, 벤쿠버섬 마못

Size 크기	56~70cm / 3~7kg	**Lifespan** 평균수명	10~12 years
Diet 식성	Herbivore 초식	**Habitat** 서식지	Grassland, mountain 초원, 산
Behavior 행동	Group living 무리생활	**Distribution** 분포지역	Vancouver island in British Columbia 브리티쉬 컬럼비아주 벤쿠버섬
Reproduction 번식	Average 4 평균 4마리	**Conservation status** 멸종위기 등급	Critically endangered (CR) 위급종

Vancouver island marmot was designated as a "sidekick" to the official mascots -Miga, Quatchi, and Sumi in 2010 Winter Olympics and Paralympics.

벤쿠버섬 마못은 2010년 동계올림픽 및 장애인 올림필의 공식 마스코트인 수미,콰치,미가의 보조 마스코트로 활약하였습니다.

CR

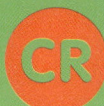

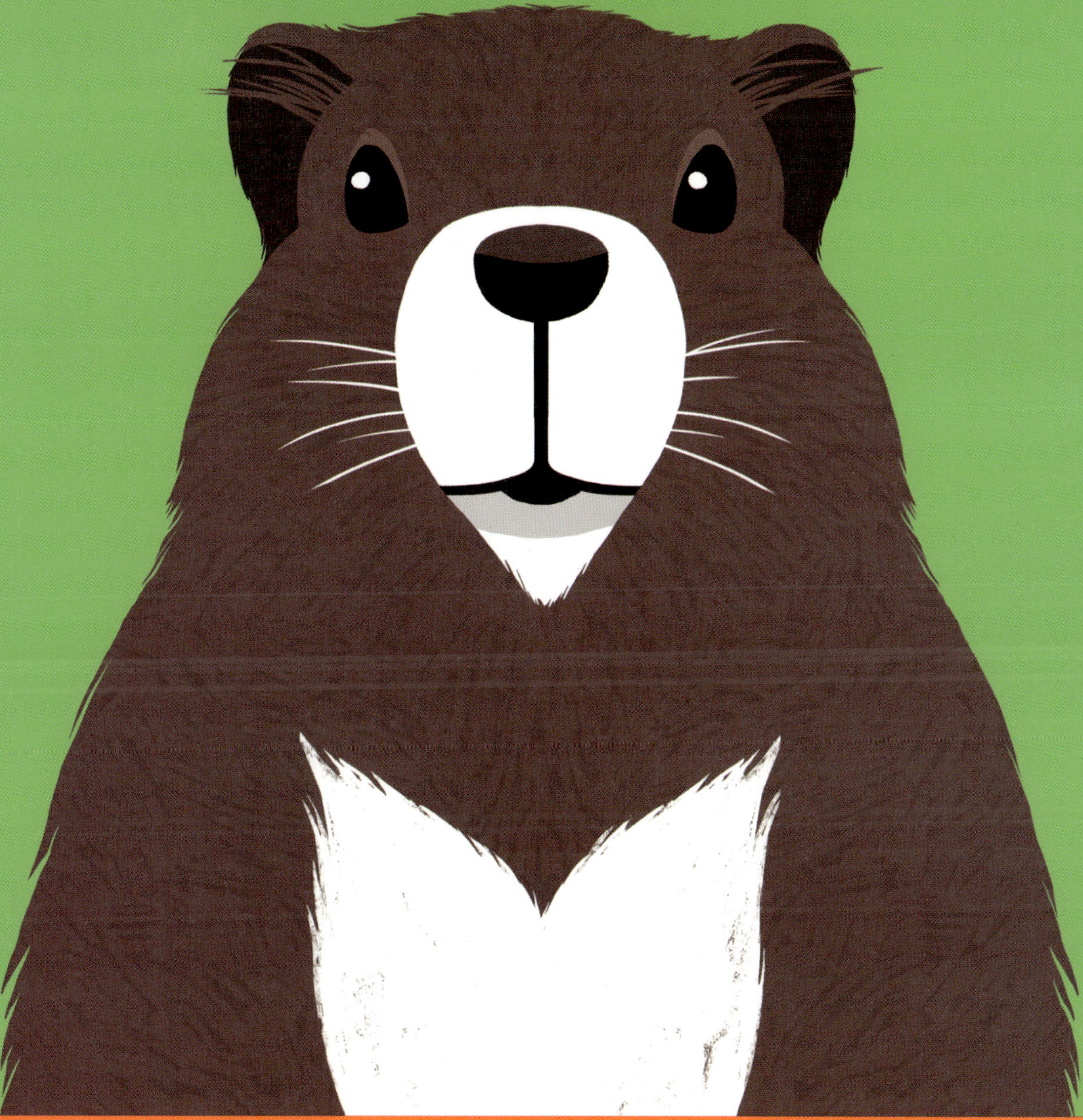

- EX Extinct
- EW Extinct in the Wild
- CR Critically Endangered
- EN Endangered
- VU Vulnerable
- NT Near Threatened
- LC Least Concern

VANCOUVER ISLAND MARMOT
CRITICALLY ENDANGERED SPECIES

Harp seal, ice lover
얼음 애호가, 하프 물범

Size 크기	1.5~2m / 120~135kg	**Lifespan** 평균수명	20~35 years
Diet 식성	Carnivore 육식	**Habitat** 서식지	Ocean 바다
Behavior 행동	Solitary living 단독생활	**Distribution** 분포지역	Arctic, northern Atlantic Ocean 북극해, 북대서양
Reproduction 번식	Average 평균 1마리	**Conservation status** 멸종위기 등급	Least concern (LC) 관심대상종

HARP SEAL
LEAST CONCERN SPECIES

EX Extinct
EW Extinct in the Wild
CR Critically Endangered
EN Endangered
VU Vulnerable
NT Near Threatened
LC Least Concern

Golden lion tamarin,
little lion on the tree
나무 위의 작은 사자, 황금사자 타마린

Size 크기	20~33cm / 0.7~0.9kg	**Lifespan** 평균수명	15~24 years
Diet 식성	Omnivore 잡식	**Habitat** 서식지	Coastal rainforest 해안 열대우림
Behavior 행동	Group living 무리생활	**Distribution** 분포지역	Southeast Brazil 브라질 남동부 지역
Reproduction 번식	Average 2 평균 2마리	**Conservation status** 멸종위기 등급	Endangered (EN) 위기종

GOLDEN LION TAMARIN
ENDANGERED SPECIES

EX Extinct
EW Extinct in the Wild
CR Critically Endangered
EN Endangered
VU Vulnerable
NT Near Threatened
LC Least Concern

QUOTES ABOUT ANIMALS

The greatness of a nation and its moral progress can be judged by the way its animals are treated.

한 나라의 위대함과 도덕성은 동물을 대하는 태도로 판단할 수 있다.

Mahatma Gandhi, 1869~1948
a leader of the Indian independence movement

동물에 대한 명사들의 한마디

If a man aspires towards a righteous life, his first act of abstinence is from injury to animals

올바른 삶을 추구하는 사람이라면, 가장 먼저 해야 할 행동은 동물을 다치게 하는 일을 그만 두는 것이다.

Albert Einstein, 1879~1955
a German theoretical physicist

Polar bear,
white Hulk of Arctic

북극의 하얀 헐크, 북극곰

Size 크기	180~250cm / 150~800kg	**Lifespan** 평균수명	27 years
Diet 식성	Carnivore 육식	**Habitat** 서식지	Ice of the Arctic ocean 북극해의 얼음
Behavior 행동	Solitary living 단독 생활	**Distribution** 분포지역	Northern areas of Canada, Alaska, Russia, Norway 캐나다 북부, 알래스카, 러시아, 노르웨이
Reproduction 번식	Average 2 평균 2마리	**Conservation status** 멸종위기 등급	Vulnerable (VU) 취약종

POLAR BEAR
VULNERABLE SPECIES

EX Extinct
EW Extinct in the Wild
CR Critically Endangered
EN Endangered
VU Vulnerable
NT Near Threatened
LC Least Concern

Reindeer,
a long-distance traveler
장거리 여행가, 순록

Size 크기	120~220cm / 100~310kg	**Lifespan** 평균수명	13~15 years
Diet 식성	Herbivore 초식	**Habitat** 서식지	Forests in the Arctic 북극 지역의 숲
Behavior 행동	Group living 무리 생활	**Distribution** 분포지역	Northern Europe, Russia, Greenland, Canada, Alaska 북유럽, 러시아, 그린란드, 캐나다, 알래스카
Reproduction 번식	Average 1 평균 1마리	**Conservation status** 멸종위기 등급	Vulnerable (VU) 취약종

The reindeer is the mammal that travels the farthest distance annually among terrestrial animals. Some populations of reindeer in certain regions travel up to 5,000 km each year.
지상 포유류 중 연간 가장 멀리 이동하는 동물입니다. 일부 지역 순록은 연간 최대 5,000km를 이동합니다.

EX Extinct
EW Extinct in the Wild
CR Critically Endangered
EN Endangered
VU **Vulnerable**
NT Near Threatened
LC Least Concern

REINDEER
VULNERABLE SPECIES

American badger, coyote's soulmate
코요테와의 협업, 아메리카 오소리

Size 크기	52~90cm / 4~12kg	**Lifespan** 평균수명	10~14years
Diet 식성	Carnivore 육식	**Habitat** 서식지	Savana, grassland 사바나, 초원
Behavior 행동	Solitary living 단독생활	**Distribution** 분포지역	Western, central United State, south- cerntal Canada 미국 중서부, 캐나다 중남부
Reproduction 번식	Average 3 평균 3마리	**Conservation status** 멸종위기등급	Least concern (LC) 관심대상종

American badger has a strong claw, and the coyote has a sensitive nose. These two hunt by cooperating with one another. The coyote finds the location of the prey and the american badger finds it by digging through the ground with its claws.
아메리카 오소리는 강한 발톱을, 코요테는 예민한 후각을 갖고있습니다. 이 둘은 종종 협업을 통해 사냥을 합니다.
코요테의 후각으로 먹잇감의 위치를 찾고, 아메리카 오소리의 강한 발톱으로 땅을 파내어 잡는 방식입니다.

Brown woolly monkey,
soft fluffy hair ball
복슬복슬 부드러운 털, 갈색양털원숭이

Size 크기	50~60cm / 3~10kg
Diet 식성	Omnivore 잡식
Behavior 행동	Group living 무리생활
Reproduction 번식	Average 1 평균 1마리
Lifespan 수명	7~10 years
Habitat 서식지	Tropical forest 열대우림
Distribution 분포지역	Colombia, Eeuador, Peru, Brazil 콜롬비아, 에콰도르, 페루, 브라질
Conservation status 멸종위기등급	Vulnerable (VU) 취약종

Brown woolly monkey has short thick curly hair all over its body.
갈색양털원숭이는 짧고 두꺼우며 곱슬거리는 털로 온몸이 뒤덮여 있습니다.

BROWN WOOLLY MONKEY
VULNERABLE SPECIES

EX Extinct
EW Extinct in the Wild
CR Critically Endangered
EN Endangered
VU Vulnerable
NT Near Threatened
LC Least Concern

Red fox,
the smart hunter
영리한 사냥꾼, 붉은 여우

Size 크기	45.5~90cm / 3~14kg	**Lifespan** 평균수명	4 years
Diet 식성	Omnivore 잡식	**Habitat** 서식지	Tundra, desert, mountain 툰드라, 사막, 산
Behavior 행동	Solitary living 단독 생활	**Distribution** 분포지역	North Africa, North America, Eurasia 북아프리카, 북아메리카, 유라시아
Reproduction 번식	Average 4.5 평균 4.5마리	**Conservation status** 멸종위기 등급	Least concern (LC) 관심대상종

RED FOX
LEAST CONCERN SPECIES

EX Extinct
EW Extinct in the Wild
CR Critically Endangered
EN Endangered
VU Vulnerable
NT Near Threatened
LC Least Concern

White-nosed coati,
jungle adventurer
밀림 탐험가, 흰코 코아티

Size 크기	80~130cm / 3~5kg	**Lifespan** 평균수명	7~10years
Diet 식성	Omnivore 잡식	**Habitat** 서식지	Tropical forest 열대우림
Behavior 행동	Group living 무리생활	**Distribution** 분포지역	Central America, Mexico, Colombia 중앙아메리카, 멕시코, 콜롬비아
Reproduction 번식	Average 4 평균 4마리	**Conservation status** 멸종위기등급	Least concern (LC) 관심대상종

White-nosed coati searches through the jungle more than 2,000-meter radius a day to find food.
흰코 코아티는 음식을 구하기 위해 하루에 밀림속에서 최대 반경 2,000미터를 탐험합니다.

Honduran white bat,
a small cotton ball in the forest
숲속의 작은 솜뭉치, 온두라스 흰박쥐

Size 크기	40~47cm / 4~7g	Lifespan 평균수명	7 years
Diet 식성	Omnivore 잡식	Habitat 서식지	Forest 숲
Behavior 행동	Group living 무리생활	Distribution 분포지역	Costa Rica, Honduras, Nicaragua, Panama 코스타리카, 온두라스, 니카라과, 파나마
Reproduction 번식	Average 1 평균 1마리	Conservation status 멸종위기등급	Near threatened (NT) 준위협종

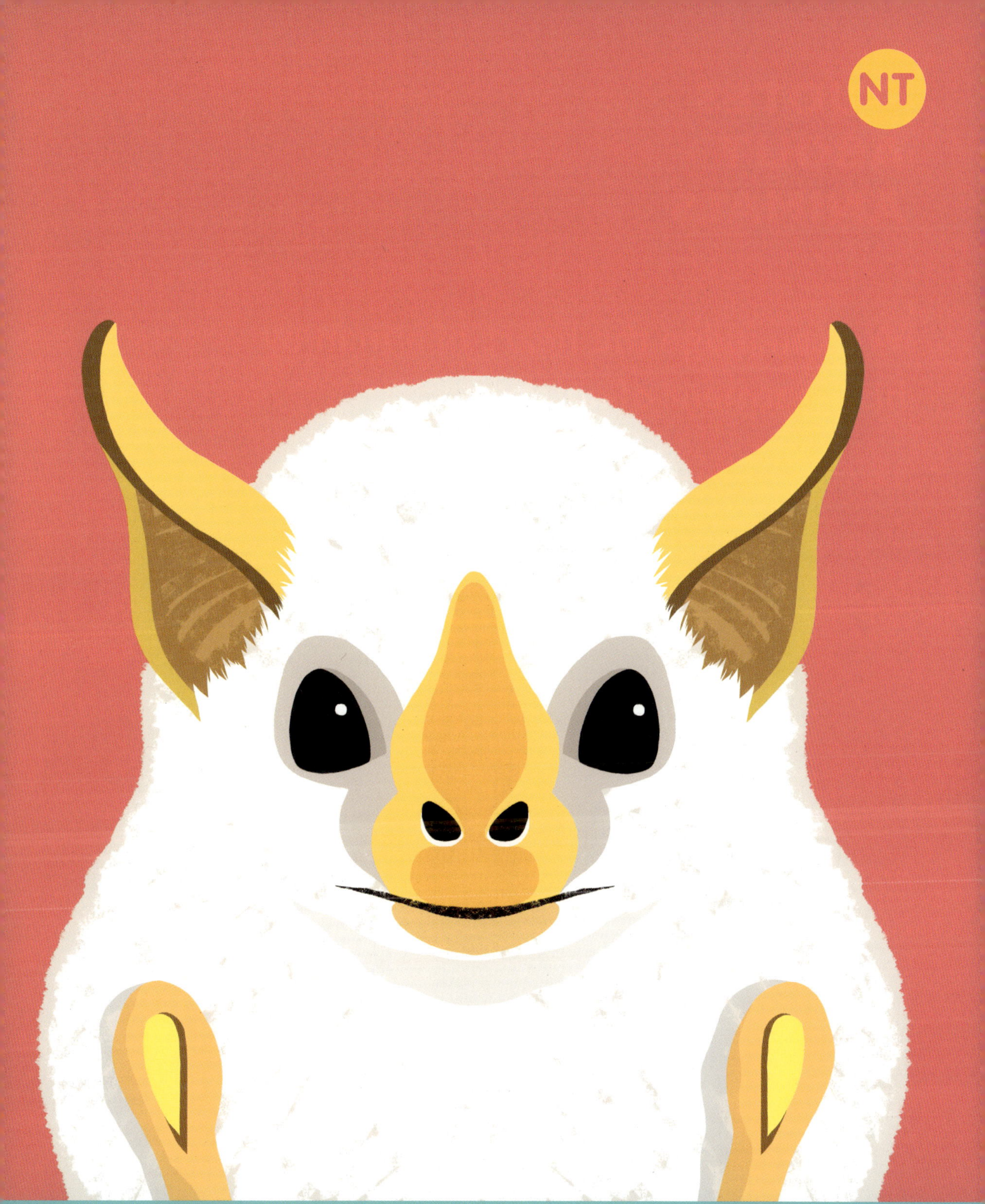

HONDURAN WHITE BAT
NEAR THREATENED SPECIES

- EX Extinct
- EW Extinct in the Wild
- CR Critically Endangered
- EN Endangered
- VU Vulnerable
- NT Near Threatened
- LC Least Concern

QUOTES ABOUT ANIMALS

We can judge the heart of a man by his treatment of animals.

동물을 대하는 태도를 보면 그 사람의 본성을 판단할 수 있다.

Immanuel Kant, 1724~1804

a German philosopher

동물에 대한 명사들의 한마디

The question is not,
"Can they reason?" nor
"Can they talk?" but rather,
"Can they suffer?"

문제는 "동물이 사고를 하느냐"도 아니고
"그들이 말할 수 있는가"도 아니다.
"그들이 고통받고 있는가"이다.

Jeremy Bentham, 1748~1832
an English philosopher, jurist

Capybara,
chilled-out diver
느긋한 잠수부, 카피바라

Size 크기	106~134cm / 35~66kg	**Lifespan** 평균수명	6~10 years
Diet 식성	Herbivore 초식	**Habitat** 서식지	Grassland 초원
Behavior 행동	Group living 무리생활	**Distribution** 분포지역	South America 남아메리카
Reproduction 번식	Average 2~8 평균 2~8마리	**Conservation status** 멸종위기등급	Least concern (LC) 관심대상종

Capybara is meek unlike its outer feature. Although it usually lives on the ground,
it can also swim and dive for 5 minutes maximum under water, due to its webbed feet.
카피바라는 큰 덩치와는 달리 온순한 성격입니다. 평소에는 육지 생활을 하지만 물갈퀴를 갖고있어 최대 5분 동안 잠수와 수영을 할 수 있습니다.

CAPYBARA
LEAST CONCERN SPECIES

EX Extinct
EW Extinct in the Wild
CR Critically Endangered
EN Endangered
VU Vulnerable
NT Near Threatened
LC Least Concern

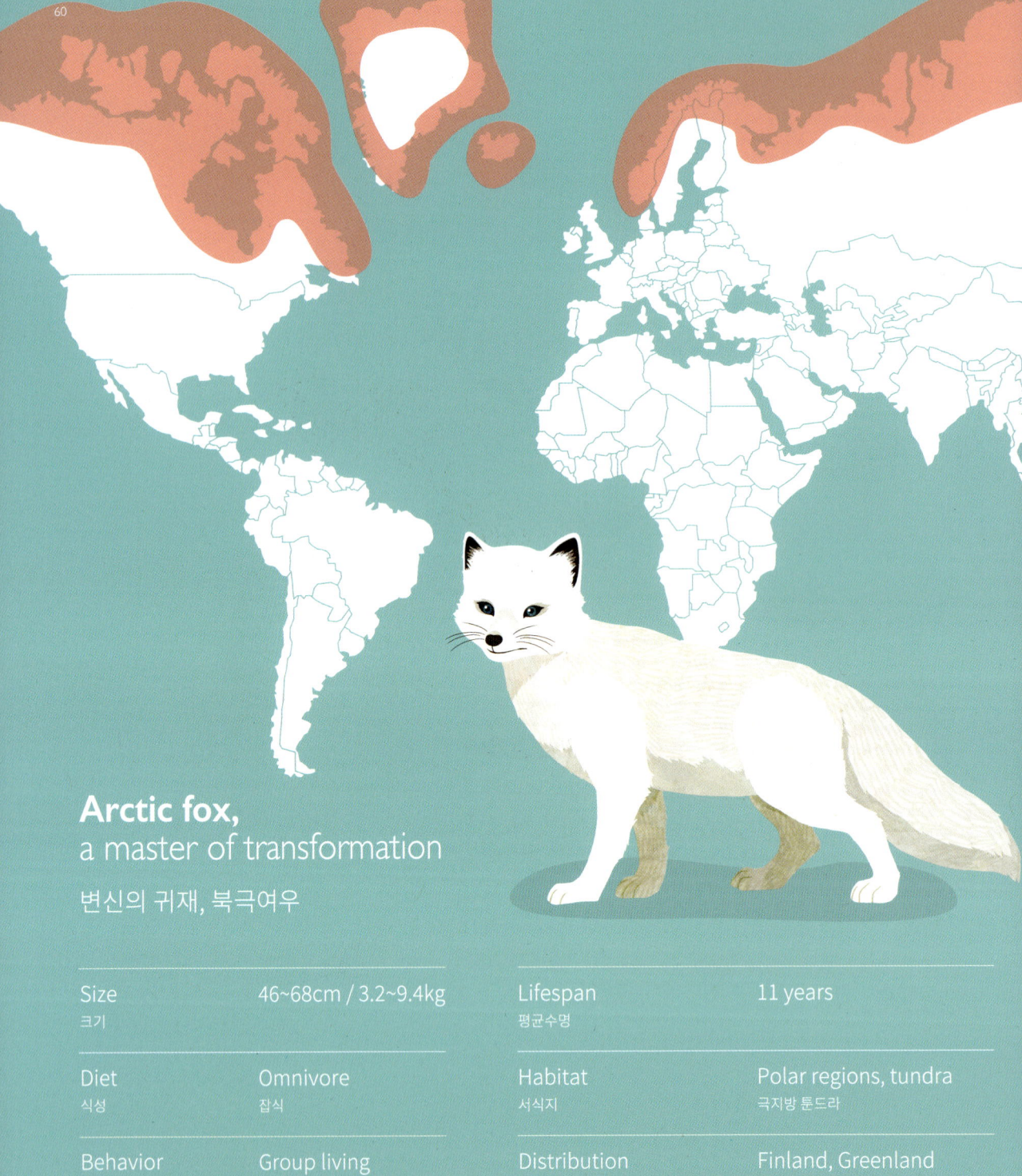

Arctic fox,
a master of transformation
변신의 귀재, 북극여우

Size 크기	46~68cm / 3.2~9.4kg	**Lifespan** 평균수명	11 years
Diet 식성	Omnivore 잡식	**Habitat** 서식지	Polar regions, tundra 극지방 툰드라
Behavior 행동	Group living 무리생활	**Distribution** 분포지역	Finland, Greenland 핀란드, 그린란드 및 북유럽, 북아시아
Reproduction 번식	Average 1~5 평균 1~5마리	**Conservation status** 멸종위기 등급	Least Concern (LC) 관심대상종

The Arctic fox is the only canid that changes its fur color through molting depending on the season, and in summer, it molts into brown and gray fur.
북극여우는 계절에 따라 털갈이를 통해 털 색깔이 변하는 유일한 개과류로 여름에는 갈색과 회색 털로 털갈이를 합니다.

ARCTIC FOX
LEAST CONCERN SPECIES

- EX Extinct
- EW Extinct in the Wild
- CR Critically Endangered
- EN Endangered
- VU Vulnerable
- NT Near Threatened
- LC Least Concern

Blonde capuchin,
the rediscovered monkey
재발견된 원숭이, 금발 카푸친

Size 크기	35~40cm / 2~3kg	**Lifespan** 평균수명	15~25 years
Diet 식성	Omnivore 잡식	**Habitat** 서식지	Forest 숲
Behavior 행동	Group living 무리생활	**Distribution** 분포지역	Northeastern Brazil 브라질 북동부 지역
Reproduction 번식	Average 1 평균 1마리	**Conservation status** 멸종위기등급	Endangered (EN) 위기종

Blonde capuchin was estimated to be extinct, but then was rediscovered in 2006.
금발 카푸친은 멸종된 것으로 추정되다가 2006년 재발견되었습니다.

BLONDE CAPUCHIN
ENDANGERED SPECIES

EX Extinct
EW Extinct in the Wild
CR Critically Endangered
EN **Endangered**
VU Vulnerable
NT Near Threatened
LC Least Concern

Mountain tapir,
elephant's nose
코끼리 코, 산악맥

Size 크기	180~250cm / 136~250kg
Diet 식성	Herbivore 초식
Behavior 행동	Solitay living 단독생활
Reproduction 번식	Average 1 평균 1마리
Lifespan 평균수명	30~35 years in captivity 사육시 30~35년
Habitat 서식지	Rinforest, mountain 열대우림, 산
Distribution 분포지역	Colombia, Ecuado, Peru 콜롬비아, 에콰도르, 페루
Conservation status 멸종위기등급	Endangered (EN) 위기종

Mountain tapir has a long enough nose to grab leafs which is what they usually feed on.
산악맥은 주 먹이인 나뭇잎을 잡을 수 있을 정도로 긴 코를 가졌습니다.

MOUNTAIN TAPIR
ENDANGERED SPECIES

- EX Extinct
- EW Extinct in the Wild
- CR Critically Endangered
- EN Endangered
- VU Vulnerable
- NT Near Threatened
- LC Least Concern

Guanaco,
Lama's family

라마 가족, 구아나코

Size 크기	190~250cm / 90~140kg
Diet 식성	Herbivore 초식
Behavior 행동	Group living 무리생활
Reproduction 번식	Average 1 평균 1마리
Lifespan 평균수명	28~33 years in captivity 사육시 28~33년
Habitat 서식지	Savana, lowland 사바나, 저지대
Distribution 분포지역	Altiplano of Peru, Chile, Bolivia, Algentian 페루의 안데스 고산 지대, 칠레, 볼리비아, 아르헨티나
Conservation status 멸종위기등급	Least concern (LC) 관심대상종

Jaguarundi,
small Jaguar in the wild
야생의 작은 재규어, 재규어런디

Size 크기	55~77cm / 3.5~9kg	Lifespan 평균수명	15 years in captivity 사육시 최대 15년
Diet 식성	Carnivore 육식	Habitat 서식지	Rainforest, grassland 열대우림, 초원
Behavior 행동	Group living 무리생활	Distribution 분포지역	Southern North America, South America 북아메리카 남쪽, 남아메리카
Reproduction 번식	Average 2~3 평균 2~3마리	Conservation status 멸종위기등급	Least concern (LC) 관심대상종

Jaguarundi is the only wild cat that has a single colored hair.
재규어런디는 야생고양이 중 유일하게 단색의 털을 가지고있습니다.

Weddell seal,
big chubby sweetie
덩치 큰 순둥이, 웨델 바다표범

Size 크기	250~350cm / 400~600kg	**Lifespan** 평균수명	18 years
Diet 식성	Crustaceans, fish 갑각류, 어류	**Habitat** 서식지	Waterfront 해안가
Behavior 행동	Group living 무리생활	**Distribution** 분포지역	Antarctic regions 남극지방
Reproduction 번식	Average 1 평균 1마리	**Conservation status** 멸종위기 등급	Least concern (LC) 관심대상종

A Weddell seal has big eyes, a chubby body, and a mild personality.
웨델 바다표범은 큰 눈과 통통한 몸매이고 성격이 유순한 것이 특징입니다.

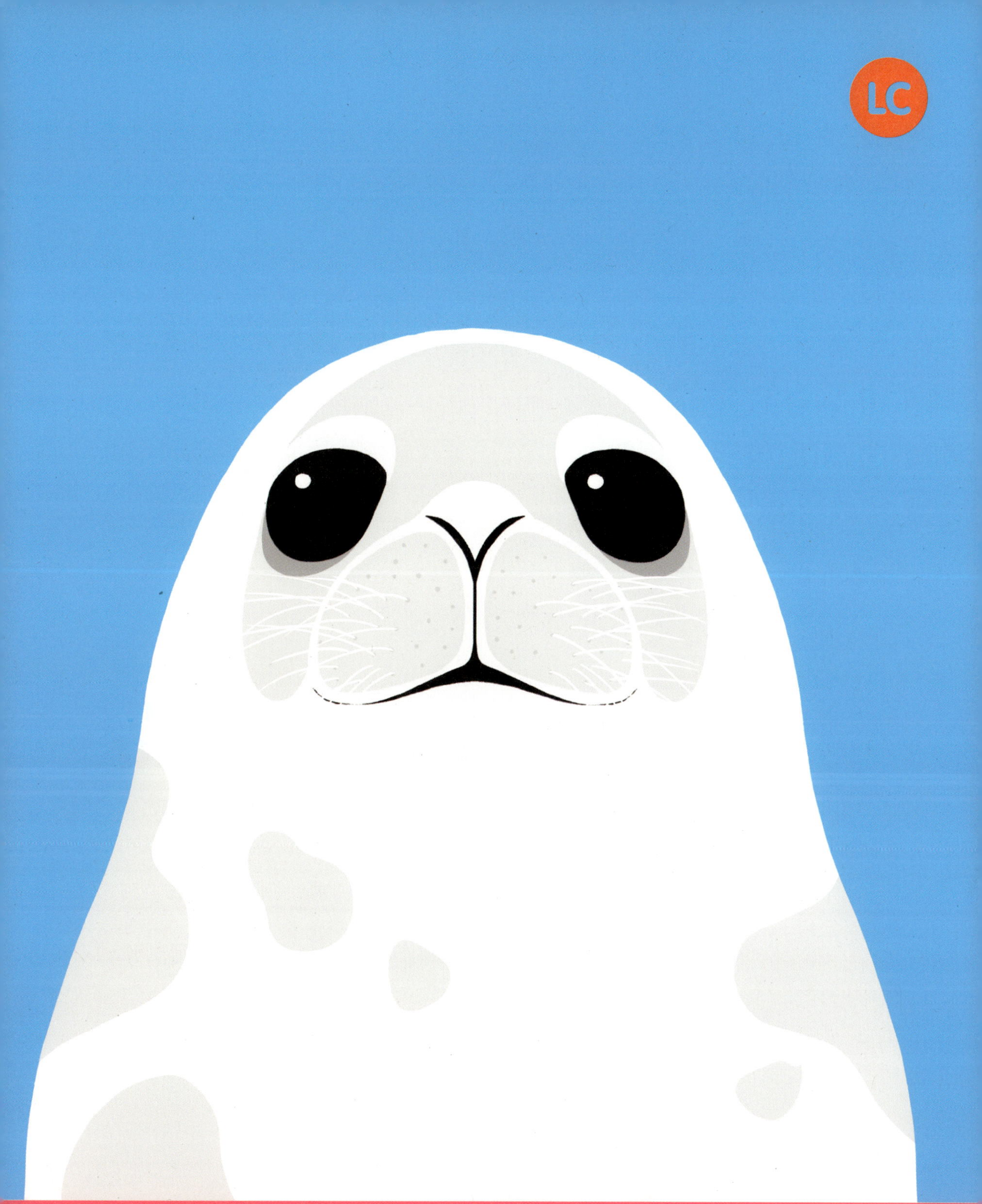

WEDDELL SEAL
LEAST CONCERN SPECIES

- EX Extinct
- EW Extinct in the Wild
- CR Critically Endangered
- EN Endangered
- VU Vulnerable
- NT Near Threatened
- LC Least Concern

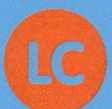

QUOTES ABOUT ANIMALS

It is just like man's vanity and impertinence to call an animal dumb because it is dumb to his dull perceptions.

사람이 동물을 멍청하다고 부르는 것은
그 사람의 자만심과 건방짐 때문이며,
자신의 아둔함으로 인해 동물이 멍청해 보일 뿐이다.

Mark Twain, 1835~1910
an American writer

동물에 대한 명사들의 한마디

Atrocities are not less atrocities when they occur in laboratories and are called medical research.

의학 연구라는 명목으로 동물 실험을 한다고 해서 잔혹함이 줄어드는 것이 아니다.

George Bernard Shaw, 1856~1950
an Irish playwright

Gray-handed night monkey,
a father full of paternal love
부성애 넘치는 회색손 올빼미 원숭이

Size 크기	44~48cm / 0.9~1kg	**Lifespan** 평균수명	33 years in captivity 사육시 최대 33년
Diet 식성	Omnivore 잡식	**Habitat** 서식지	Forest 숲
Behavior 행동	Group living 무리생활	**Distribution** 분포지역	Part of Colombia 일부 콜롬비아 지역
Reproduction 번식	Average 1 평균 1마리	**Conservation status** 멸종위기등급	Vulnerable 취약종

The male carries the infant from the time it is one or two days old, passing it to the female for nursing.
회색손 올빼미 원숭이는 새끼를 낳은 후 젖을 먹이기 위해 어미에게 넘겨진 다음, 하루 이틀 후 부터는 수컷이 새끼를 기릅니다.

GRAY-HANDED NIGHT MONKEY
VULNERABLE SPECIES

- EX Extinct
- EW Extinct in the Wild
- CR Critically Endangered
- EN Endangered
- VU **Vulnerable**
- NT Near Threatened
- LC Least Concern

Bald uakari,
an eye-catching face
시선을 사로잡는 얼굴, 대머리 우아카리

Size 크기	45~57cm / 2.7-3.4kg	**Lifespan** 평균수명	10 years
Diet 식성	Herbivore 초식	**Habitat** 서식지	Forest 숲
Behavior 행동	Group living 무리 생활	**Distribution** 분포지역	Brazil, Peru 브라질, 페루
Reproduction 번식	Average 1 평균 1마리	**Conservation status** 멸종위기 등급	Vulnerable (VU) 취약종

The red facial skin of a Bald Uakari is considered as an indicator of health, and it receive selection as a healthy mate.
붉은 얼굴 피부는 건강의 척도이며, 건강한 짝의 선택을 받을 수 있습니다.

BALD UAKARI
VULNERABLE SPECIES

EX Extinct
EW Extinct in the Wild
CR Critically Endangered
EN Endangered
VU Vulnerable
NT Near Threatened
LC Least Concern

Eastern Spotted Skunk,
smaller and more dainty than a squirrel

다람쥐보다도 작고 갸냘픈, 동부 얼룩 스컹크

Size 크기	24~26cm / 0.2-1.8kg	Lifespan 평균수명	20 years
Diet 식성	Omnivore 잡식	Habitat 서식지	Forest, Grassland 숲, 초원
Behavior 행동	Solitary living 단독 생활	Distribution 분포지역	Eastern USA, Canada, Mexico 미 동부 내륙, 캐나다, 멕시코
Reproduction 번식	Average 4~5 평균 4~5마리	Conservation status 멸종위기 등급	Vulnerable (VU) 취약종

EASTERN SPOTTED SKUNK
VULNERABLE SPECIES

EX Extinct
EW Extinct in the Wild
CR Critically Endangered
EN Endangered
VU Vulnerable
NT Near Threatened
LC Least Concern

Spectacled bear,
one and only bear in South America
남미 유일의 곰, 안경곰

Size 크기	150~200cm / 130~200kg
Diet 식성	Omnivore 잡식
Behavior 행동	Solitary living 단독 생활
Reproduction 번식	Average 2 평균 2마리
Lifespan 평균수명	20~40 years
Habitat 서식지	Coastal forest 해안 산림
Distribution 분포지역	Northwest Argentina 아르헨티나 북서부 지역
Conservation status 멸종위기등급	Vulnerable (VU) 취약종

Spectacled bear is the only ursine animal living in South America.
안경곰은 남아메리카에 사는 유일한 곰과 동물입니다.

Silky Anteater,
world smallest anteater
가장 작은 개미핥기, 애기 개미핥기

Size 크기	15-23cm / 0.2~0.4kg	**Lifespan** 평균수명	4 years
Diet 식성	Omnivore 잡식	**Habitat** 서식지	Tundra, desert, mountain 툰드라, 사막, 산
Behavior 행동	Solitary living 단독 생활	**Distribution** 분포지역	North Africa, North America, Eurasia 북아프리카, 북아메리카, 유라시아
Reproduction 번식	Average 1 평균 1마리	**Conservation status** 멸종위기 등급	Least concern (LC) 관심대상종

SILKY ANTEATER
LEAST CONCERN SPECIES

EX Extinct
EW Extinct in the Wild
CR Critically Endangered
EN Endangered
VU Vulnerable
NT Near Threatened
LC Least Concern

Andean Cat,
fearless wildcat

겁 없는 야생 고양이, 안데스 고양이

Size 크기	57-85cm / 4~ 5.5kg
Diet 식성	Omnivore 잡식
Behavior 행동	Solitary living 단독 생활
Reproduction 번식	Average 1~2 평균 1~2마리
Lifespan 평균수명	16 years
Habitat 서식지	Shrubland, Grassland, Rocky areas 관목 지대, 초원, 바위
Distribution 분포지역	Argentina, Bolivia 아르헨티나, 볼리비아
Conservation status 멸종위기등급	Endangered (EN) 위기종

The Andean cat is only found in high altitudes ranging from about 1,800 to 4,000 meters in the Andes Mountains.
안데스 산맥의 해발 1,800m~4,000m정도의 높은 곳에서만 서식합니다.

ANDEAN CAT
ENDANGERED SPECIES

EX Extinct
EW Extinct in the Wild
CR Critically Endangered
EN Endangered
VU Vulnerable
NT Near Threatened
LC Least Concern

Ringtail,
I'm not thirsty
물없이도 살아요, 알락꼬리 고양이

Size 크기	30~42cm / 0.7~1.5kg	Lifespan 평균수명	7 years
Diet 식성	Omnivore 잡식	Habitat 서식지	Grassland, Rocky areas, Forest, Shrubland 초원, 바위 지역, 숲, 관목지대,
Behavior 행동	Solitary living 단독 생활	Distribution 분포지역	Canada, Mexico, U.S.A. 캐나다, 멕시코, 미국
Reproduction 번식	Average 2~4 평균 2~4마리	Conservation status 멸종위기 등급	Least concern (LC) 관심대상종

Ringtails living in dry areas can survive for long periods with just the water they obtain from their food.
건조한 지역에 사는 링테일은 음식을 섭취해 얻은 물만으로도 오랜 기간 생존할 수 있습니다.

QUOTES ABOUT ANIMALS

As long as man continues to be
the ruthless destroyer of lower living beings,
he will never know health or peace.
For as long as men massacre animals,
they will kill each other.

동물들을 잔인하게 죽이는 한,
인류는 건강과 평화를 결코 알지 못할 것이다.
사람들이 동물을 학살하는 한,
그들은 서로 죽이는 것을 멈추지 않을 것이다.

Pythagoras, B.C.580 ~ B.C.500
an Ionian Greek mathematician

동물에 대한
명사들의
한마디

I am in favor of animal rights
as well as human rights.
That is the way of a whole human being.

나는 인권만큼이나 동물의 권익을 소중히 여긴다.
이는 모든 인류가 나아가야 할 길이다.

Abraham Lincoln, 1809~1865
The 16th President of the United States

Island fox,
smallest fox in U.S.A.
미국에서 가장 작은 여우, 아일랜드 여우

Size 크기	48-50cm / 1~2.8kg
Diet 식성	Omnivore 잡식
Behavior 행동	Solitary living 단독생활
Reproduction 번식	Average 2~3 평균 2~3마리
Lifespan 평균수명	8 years in captivity 사육시 평균 8년
Habitat 서식지	Forest, Grassland 숲, 초원
Distribution 분포지역	Islands of California 캘리포니아의 섬 지역
Conservation status 멸종위기등급	Near Threatened (NT) 준위협종

ISLAND FOX
NEAR THREATENED SPECIES

- EX Extinct
- EW Extinct in the Wild
- CR Critically Endangered
- EN Endangered
- VU Vulnerable
- NT Near Threatened
- LC Least Concern

Bighorn Sheep,
the majestic male of the grassland
초원의 우람한 수컷, 큰뿔양

Size 크기	90~105cm / 60~143kg	**Lifespan** 평균수명	7~8 years
Diet 식성	Herbivore 초식	**Habitat** 서식지	Grassland, Rocky areas, Shrubland 초원, 바위 지역, 관목지대
Behavior 행동	Group living 무리 생활	**Distribution** 분포지역	Canada, Mexico, U.S.A. 캐나다, 멕시코, 미국
Reproduction 번식	Average 1~2 평균 1~2마리	**Conservation status** 멸종위기 등급	Least concern (LC) 관심대상종

Male bighorn sheep are nearly twice as large in size as females.
큰뿔양은 수컷이 암컷보다 2배 가까이 덩치가 큽니다.

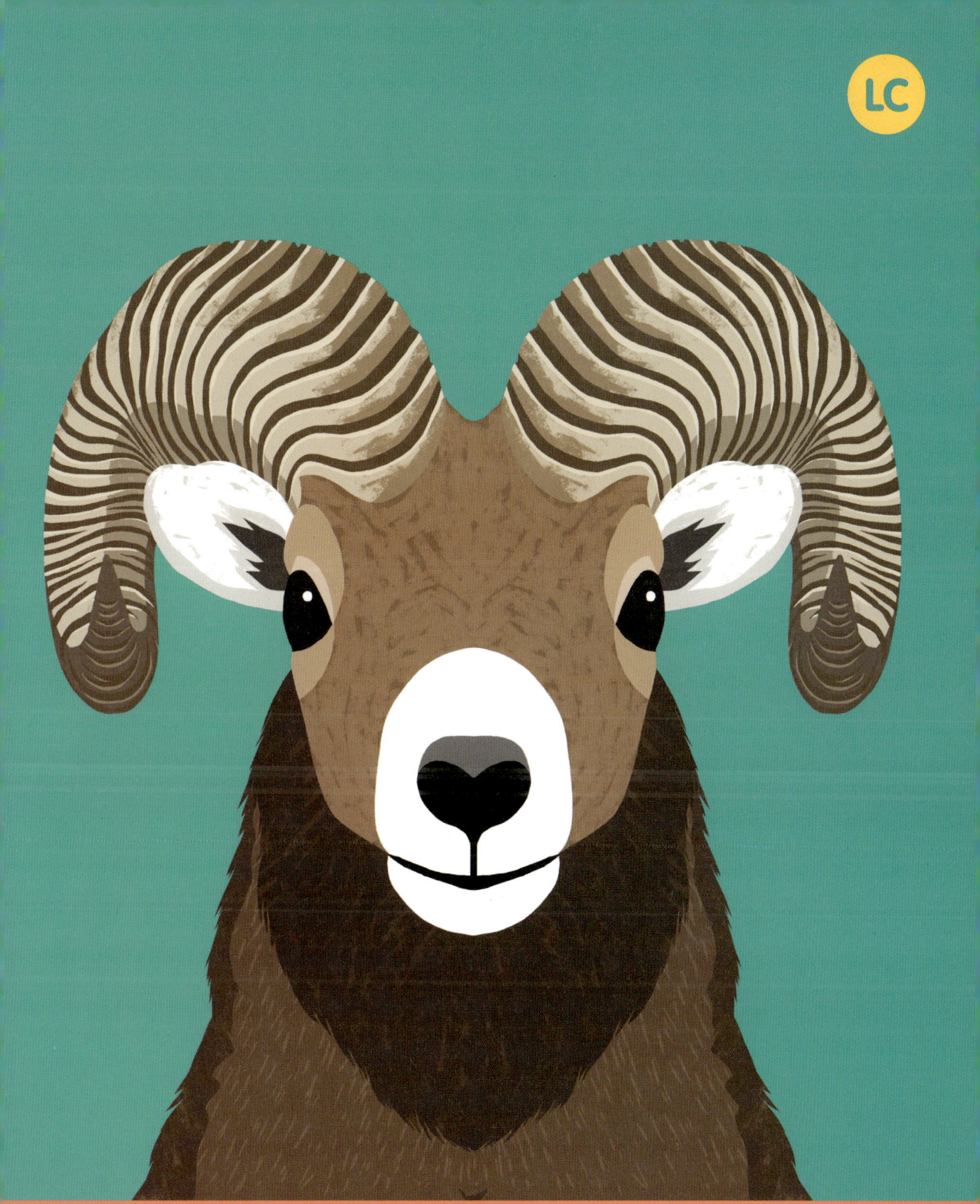

- EX Extinct
- EW Extinct in the Wild
- CR Critically Endangered
- EN Endangered
- VU Vulnerable
- NT Near Threatened
- LC Least Concern

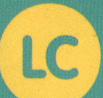

BIGHORN SHEEP
LEAST CONCERN SPECIES

Tehuantepec Jackrabbit,
the largest wild rabbit
가장 큰 야생산토끼, 테우안테펙산토끼

Size 크기	53~61cm / 3.5~4kg	**Lifespan** 평균수명	2.8 years
Diet 식성	Herbivore 초식	**Habitat** 서식지	Savana, Shrubland, Grassland 사바나, 관목지대, 초원
Behavior 행동	Group living 무리 생활	**Distribution** 분포지역	Mexico 멕시코
Reproduction 번식	Average 1~4 평균 1~4마리	**Conservation status** 멸종위기 등급	Endangered (EN) 위기종

Galapagos fur seal,
I'm not sea-lion

바다사자가 아니에요, 갈라파고스 물개

Size 크기	125~160cm / 30~64kg	**Lifespan** 평균수명	22 years
Diet 식성	Fish, squid, shellfish 어류, 오징어, 조개류	**Habitat** 서식지	Marine Neritic 얕은 바다
Behavior 행동	Group living 무리 생활	**Distribution** 분포지역	Galápagos Islands of Ecuador 에콰도르 갈라파고스섬
Reproduction 번식	Average 1 평균 1마리	**Conservation status** 멸종위기 등급	Endangered (EN) 위기종

Galapagos fur seals look very similar to sea lions, making it difficult to distinguish them without a proper understanding of their distinguishing characteristics. 갈라파고스 물개는 바다사자와 매우 비슷하게 생겨서 특징을 제대로 알지 못하면 구분하기 어렵습니다.

GALAPAGOS FUR SEAL
ENDANGERED SPECIES

EX Extinct
EW Extinct in the Wild
CR Critically Endangered
EN Endangered
VU Vulnerable
NT Near Threatened
LC Least Concern

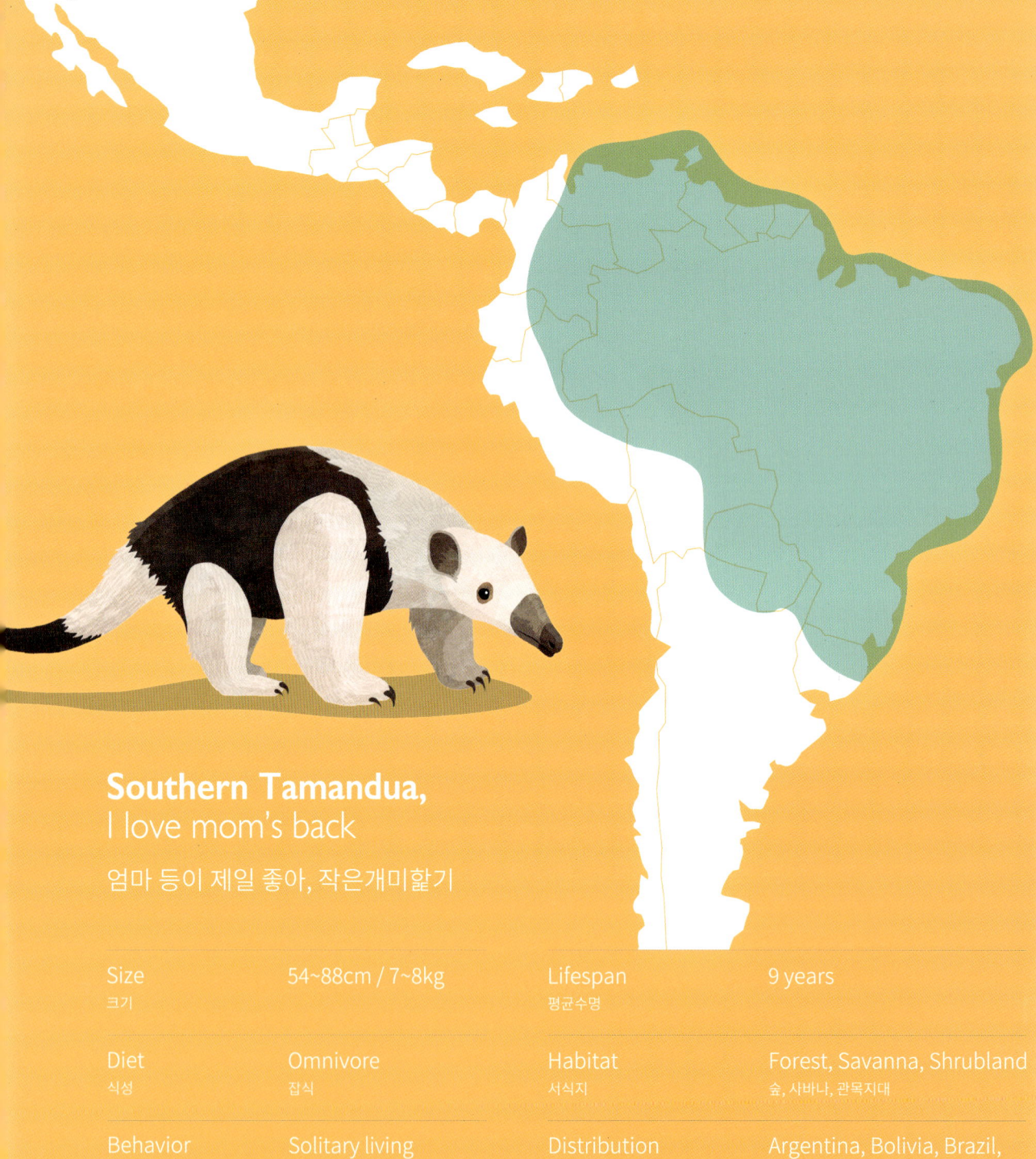

Southern Tamandua,
I love mom's back

엄마 등이 제일 좋아, 작은개미핥기

Size 크기	54~88cm / 7~8kg	**Lifespan** 평균수명	9 years
Diet 식성	Omnivore 잡식	**Habitat** 서식지	Forest, Savanna, Shrubland 숲, 사바나, 관목지대
Behavior 행동	Solitary living 단독생활	**Distribution** 분포지역	Argentina, Bolivia, Brazil, Venezuela 아르헨티나, 볼리비아, 브라질, 베네수엘라
Reproduction 번식	Average 1 평균 1마리	**Conservation status** 멸종위기 등급	Least concern (LC) 관심대상종

It rides on the mother's back for several months up to a year.
수개월에서 최대 1년 동안 어미의 등에 올라타 생활합니다.

SOUTHERN TAMANDUA
LEAST CONCERN SPECIES

- **EX** Extinct
- **EW** Extinct in the Wild
- **CR** Critically Endangered
- **EN** Endangered
- **VU** Vulnerable
- **NT** Near Threatened
- **LC** Least Concern

Maned Wolf,
I'm not skunk
스컹크는 아닙니다, 갈기늑대

Size 크기	87~130cm / 20~25kg	**Lifespan** 평균수명	12~15 years
Diet 식성	Carnivore 육식	**Habitat** 서식지	Forest, Savanna, Shrubland 숲, 사바나, 관목지대
Behavior 행동	Solitary living 단독생활	**Distribution** 분포지역	Northern Bolivia, Southern Brazil, Eastern Andes 브라질 남부, 안데스산맥 동부, 볼리비아 북부
Reproduction 번식	Average 1~5 평균 1~5마리	**Conservation status** 멸종위기 등급	Near threatened (NT) 준위협종

A Maned Wolf emits a unique marijuana-like odor when marking its territory and marks locations where it has found food with urine scent.
갈기늑대는 영역 표시를 할 때 독특한 대마초 냄새를 풍기고 먹이가 있는 장소를 소변 냄새로 표시해 둡니다.

MANED WOLF
NEAR THREATENED SPECIES

EX Extinct
EW Extinct in the Wild
CR Critically Endangered
EN Endangered
VU Vulnerable
NT Near Threatened
LC Least Concern

Blue-throated Macaw,
eye-transformer

변신의 눈동자, 푸른목 금강앵무

Size 크기	85-90cm / 0.75-1.1kg
Diet 식성	Herbivore 초식
Behavior 행동	Group living 무리생활
Reproduction 번식	Average 1~3 평균 1~3마리
Lifespan 평균수명	8 years
Habitat 서식지	Forest, Savanna 숲, 사바나
Distribution 분포지역	Bolivia 볼리비아
Conservation status 멸종위기등급	Critically endangered (CR) 위급종

Blue-throated macaw have gray irises when they are between 1~3 years old, but they gradually turn yellow as they age.
푸른목금강앵무는 1살~3살 때 눈동자가 회색이지만 나이가 들수록 노랗게 변합니다.

ANIMAL PROTECTION LAWS BY COUNTRY : UNITED KINGDOM

국가별 동물보호법 : 영국

왕립 동물학대 방지 협회 (RSPCA:Royal Society for the Prevention of Cruelty to Animals)는 1824년 설립된 세계에서 가장 오래된 민간동물 단체입니다. 사회적 기여를 인정받아 1840년 빅토리아 여왕으로부터 '왕립' 칭호를 받습니다. 전역에 약 200개의 지부를 두고 24시간 동물 학대 상담 전화를 운영하며 전문적인 훈련을 받은 300여 명의 RSPCA 수사관들이 일반시민들에게 제보를 받고 경찰관과 함께 현장 출동하여 개선 지도를 하거나 때에 따라 학대자를 구속하여 처벌하는 시스템이 구축되어있습니다.

RSPCA 외에도 정부 지원 없이 기부금으로 운영되는 대표적 민간 동물복지 단체 중 하나인 Mayhew Animal Home은 1886년 런던 켄설그린에 지역 커뮤니티를 기반으로 위기에 처한 동물 구조, 길고양이 중성화 수술, 입양센터 운영, 교육 활동 등 반려동물 중심의 동물복지 활동을 펼치고 있습니다.

RSPCA is the oldest private animal organization in the world. Acknowledged its social contribution by QueenVictoria in 1840, it was granted its royal status by QueenVictoria herself. RSPCA has about 200 branches all over UK & runs 24hour animal harassment consulting callcenter. They also have about 300 professionally trained investigators who work with the police in case of a call and they can also arrest and punish the offender.

Aside from RSPCA, Mayhew Animal Home, an NGO that is run solely by donation without any government's support, has been working on animal welfare mainly focusing on companion animals. They've been rescuing animals in danger, neutering stray cats, running an adoption center, educating and so on. They are actively working based on its local community, ever since it was established in Kensel Green, London, in 1886.

ANIMAL PROTECTION LAWS BY COUNTRY: FRANCE

국가별 동물보호법: 프랑스

1850년 최초로 동물 학대를 처벌하는 동물 보호법이 제정되었습니다. 1976년에는 동물도 사람처럼 지각력 있음을 인정하는 동물복지법이 만들어졌습니다.

대표적인 동물보호단체로 S.P.A(Society for the Protection of Animals)가 있습니다. S.P.A는 전국 56개의 동물 피난처와 10여개의 무료 진료소를 운영합니다. 주된 활동은 유기 동물의 주인 찾아주기, 불법 농장으로부터의 동물 구출, 경제적으로 어려움을 겪는 동물 주인에게 재정적 지원하기 등 입니다. 이외에도 전국적으로 유기동물들을 총괄 관리하고 재입양을 위한 다양한 행사들을 마련합니다.

In 1850, France was the first to legislate animal protection, punishing animal harassment. In 1976, animal welfare legislation that accepts that animals have the same recognition as human, was also passed.

SPA(Society for Protection of Animals) is one of the most well-known animal protection organization. SPA runs 56 animal shelters and 10 free clinics all over France. Their main activities are finding the owner of a lost animal, rescuing animals from illegal farms, financially supporting the animal's owner in economic crisis and so on. They also manage abandoned animals and host events for adoption nationally.

Western Pygmy Marmoset,
world smallest monkey

가장 작은 원숭이, 서부 피그미 마모셋

Size 크기	12~13cm / 0.1~0.2kg	**Lifespan** 평균수명	6~7 years
Diet 식성	Herbivore 초식	**Habitat** 서식지	Forest 숲
Behavior 행동	Group living 무리생활	**Distribution** 분포지역	Brazil, Peru, Colombia, Ecuador 브라질, 페루, 콜롬비아, 에콰도르
Reproduction 번식	Average 1~2 평균 1~2마리	**Conservation status** 멸종위기 등급	Vulnerable (VU) 취약약종

Western pygmy marmoset is the smallest monkey in the world, with an average weight of 110g for males and 122g for females.
수컷 평균 무게는 110g, 암컷 평균 무게는 122g으로 세계에서 가장 작은 원숭이입니다.

Caqueta Titi Monkey,
I'm family man

패밀리맨, 카케타 티티 원숭이

Size 크기	35cm / 0.8~1.4kg	Lifespan 평균수명	25 years in captivity 사육시 평균 25년
Diet 식성	Omnivore 잡식	Habitat 서식지	Forest 숲
Behavior 행동	Group living 무리생활	Distribution 분포지역	Colombia 콜롬비아
Reproduction 번식	Average 1~2 평균 1~2마리	Conservation status 멸종위기 등급	Critically endangered (CR) 위급종

The titi monkey mate with only one female for life, and they have offspring to create a family.
Until the offspring reach a certain size, they spend most of their time being carried by the male.
평생 한 암컷과만 짝짓기를 하고 새끼를 낳아 무리를 만듭니다. 새끼가 크기 전까지 대부분의 시간을 수컷이 안고 생활합니다.

CAQUETA TITI MONKEY
CRITICALLY ENDANGERED SPECIES

EX Extinct
EW Extinct in the Wild
CR Critically Endangered
EN Endangered
VU Vulnerable
NT Near Threatened
LC Least Concern

American marten, hunter in the dawn
황혼의 사냥꾼, 아메리카 담비

Size 크기	32~45cm / 0.3~1.3kg	**Lifespan** 평균수명	Up to 17 years in captivity 사육시 최대 17년
Diet 식성	Carnivore 육식	**Habitat** 서식지	Coniferous forest 침엽수림
Behavior 행동	Solitary living 단독생활	**Distribution** 분포지역	Northern north America 북아메리카 북부지역
Reproduction 번식	Average 3 평균 3마리	**Conservation status** 멸종위기등급	Least concern (LC) 관심대상종

American marten mostly hunts in the dawn.
아메리카 담비는 새벽과 황혼 시간대에 가장 활발하게 사냥합니다.

Barn owl,
ghost-masked
유령의 가면을 쓴, 원숭이 올빼미

Size 크기	32~40cm / 0.4~0.6kg	**Lifespan** 평균수명	20~34 years
Diet 식성	Carnivore 육식	**Habitat** 서식지	Grassland, forest 초원, 숲
Behavior 행동	Solitary living 단독생활	**Distribution** 분포지역	Most of continent except Antarctica 남극대륙을 제외한 대부분의 대륙
Reproduction 번식	Average 4~7eggs 평균 4~7개의 알	**Conservation status** 멸종위기등급	Least concern (LC) 관심대상종

Barn owl is called ghost owl or masked owl because of its white face and feather.
원숭이 올빼미는 하얀 얼굴과 깃털 때문에 유령 올빼미 또는 가면 올빼미라고 불렸습니다.

BARN OWL
LEAST CONCERN SPECIES

EX Extinct
EW Extinct in the Wild
CR Critically Endangered
EN Endangered
VU Vulnerable
NT Near Threatened
LC Least Concern

Sea otter,
master of backstrokes
배영의 달인, 해달

Size 크기	70~120cm / 15~45kg	Lifespan 평균수명	12~23 years
Diet 식성	Carnivore 육식	Habitat 서식지	Offshore 육지와 가까운 바다
Behavior 행동	Group living 무리생활	Distribution 분포지역	Coasts of the northern and eastern North Pacific Ocean 북태평양 북부와 동부 해안
Reproduction 번식	Average 1 평균 1마리	Conservation status 멸종위기등급	Endangered (EN) 위기종

SEA OTTER
ENDANGERED SPECIES

- EX Extinct
- EW Extinct in the Wild
- CR Critically Endangered
- EN Endangered
- VU Vulnerable
- NT Near Threatened
- LC Least Concern

Squirrel monkey,
naughty in jungle
정글의 장난꾸러기, 다람쥐 원숭이

Size 크기	22~30cm / 11.5~25kg	**Lifespan** 평균수명	17 years
Diet 식성	Omnivore 잡식	**Habitat** 서식지	Rainforest 열대우림
Behavior 행동	Group living 무리생활	**Distribution** 분포지역	Panama and Costa Rica 파나마, 코스타리카
Reproduction 번식	Average 1 평균 1마리	**Conservation status** 멸종위기 등급	Least concern (LC) 관심대상종

A squirrel monkey has a voracious curiosity and likes to play pranks.
In the wild, it watches humans from nearby trees.
호기심이 왕성하고 장난을 좋아합니다. 야생에서도 사람을 보면 나무 위에서 관찰하듯이 내려다봅니다.

SQUIRREL MONKEY
LEAST CONCERN SPECIES

EX Extinct
EW Extinct in the Wild
CR Critically Endangered
EN Endangered
VU Vulnerable
NT Near Threatened
LC Least Concern

Mandrill,
shaman of 'Lion King'
라이온 킹의 주술사, 맨드릴

Size 크기	61~76cm / 11~25kg
Diet 식성	Omnivore 잡식
Behavior 행동	Group living 무리 생활
Reproduction 번식	Average 1 평균 1마리
Lifespan 평균수명	46 years in captivity 사육시 평균 46년
Habitat 서식지	Rainforest 열대우림
Distribution 분포지역	Cameroon, Gabon, Congo Equatorial Guinea 카메룬, 가봉, 콩고, 적도 기니
Conservation status 멸종위기 등급	Vulnerable (VU) 취약종

Bateleur,
street performer
곡예꾼, 달마수리

Size 크기	80~85cm / 1.8~2.7kg
Diet 식성	Small mammal 작은 포유동물
Behavior 행동	Solitary living 단독 생활
Reproduction 번식	Average 1 평균 1마리
Lifespan 평균수명	22.5 years
Habitat 서식지	Savana 사바나
Distribution 분포지역	Savana country within Sub-saharan Africa 사하라 이남 아프리카의 사바나 국가
Conservation status 멸종위기 등급	Endangered (EN) 위기종

A bateleur is called a street performer because it looks like it is juggling on a rope when it flies.
비행하는 모습이 줄을 타며 곡예하는것처럼 보여 곡예비행사라 불립니다.

BATELEUR
ENDANGERED SPECIES

- EX Extinct
- EW Extinct in the Wild
- CR Critically Endangered
- EN Endangered
- VU Vulnerable
- NT Near Threatened
- LC Least Concern

Cheetah,
the fastest mammalia
네 발의 우사인 볼트, 치타

Size 크기	112~150cm / 21~72kg	**Lifespan** 평균수명	7 years
Diet 식성	Carnivore 육식	**Habitat** 서식지	Savanna, semidesert 사바나, 반사막
Behavior 행동	Solitary living 단독 생활	**Distribution** 분포지역	The arabian peninsula to Tajikistan, Central India 아라비아 반도~타지키스탄, 중앙 인도
Reproduction 번식	Average 3 평균 3마리	**Conservation status** 멸종위기 등급	Vulnerable (VU) 취약종

Thomson's gazelle,
the grace jumper

우아한 점퍼, 톰슨가젤

Size 크기	80~120cm / 15~35kg
Diet 식성	Herbivore 초식
Behavior 행동	Group living 무리생활
Reproduction 번식	Average 1 평균 1마리
Lifespan 평균수명	10 years
Habitat 서식지	Savanna, grassland 사바나, 초원
Distribution 분포지역	Kenya, Tanzania 케냐, 탄자니아
Conservation status 멸종위기등급	Least concern (LC) 관심대상종

Red-bellied lemur,
weeping man

남자의 눈물, 붉은배 여우원숭이

Size 크기	36~54cm / 2~3kg	**Lifespan** 평균수명	20~25 years
Diet 식성	Omnivore 잡식	**Habitat** 서식지	Rainforest 열대우림
Behavior 행동	Group living 무리생활	**Distribution** 분포지역	Eastern Madagascar 마다가스카르 동부
Reproduction 번식	Average 1 평균 1마리	**Conservation status** 멸종위기등급	Vulnerable (VU) 취약종

Only the male Red-bellied lemur has the white smudges of tears in the front corner of their eyes which makes it easier to distinguish their sex.
붉은배 여우원숭이는 수컷에게만 눈 앞에 하얀색 눈물자국이 있어 암수를 쉽게 구분할 수 있습니다.

Fennc fox,
advisor of the Little Prince
어린 왕자의 조언자, 사막여우

Size 크기	35~40cm / 0.8~1.5kg
Diet 식성	Omnivore 잡식
Behavior 행동	Group living 무리생활
Reproduction 번식	Average 3 평균 3마리
Lifespan 평균수명	10 years
Habitat 서식지	Desert, sand dune 모래사막, 모래언덕
Distribution 분포지역	Sahara, Morocco 사하라, 모로코
Conservation status 멸종위기 등급	Least concern (LC) 관심대상종

FENNEC FOX
LEAST CONCERN SPECIES

EX Extinct
EW Extinct in the Wild
CR Critically Endangered
EN Endangered
VU Vulnerable
NT Near Threatened
LC Least Concern

ANIMAL PROTECTION LAWS BY COUNTRY : NORWAY

국가별 동물보호법 : 노르웨이

동물을 고의로 죽게 할 경우 1년 이상 3년 이하의 징역에 처하며 반려견의 산책이 의무화 되어있습니다. 하루 3번 이상 반려견을 산책시키지 않는 것은 동물 학대로 간주하여 3,400 크로네 ~ 17,000 크로네의 벌금이 부과됩니다. 정부에서는 동물 학대 신고 웹사이트를 개설하여 신고를 쉽게 할 수 있도록 장려하고 있습니다.

일부지역에는 동물 경찰이 별도로 존재해 조사관, 법 전문가, 코디네이터가 한 팀을 이루어 활동합니다. 이들은 모두 경찰 신분으로 동물과 관련된 사건만을 담당하여 처리합니다.

If one intentionally kills an animal, one can be sentenced from 1 to 3 years in prison and walking the companion dog is mandatory. Walking a dog less than 3 times a day is considered as animal harassment and can be fin-ed from 3400 to 17,000 krone. The government is enco-uraging reports by making a website for it.

In certain areas, there is an animal police and they work as a team with an investigator, a law professional, and a coordinator. They all are considered as official police and only handle animal related cases.

ANIMAL PROTECTION LAWS BY COUNTRY : GERMANY

국가별 동물보호법 : 독일

독일의 동물보호법 1조 1항에는 '동물과 인간은 이 세상의 동등한 창조물 이다. 누구든 합리적인 이유없이 동물에게 해를 끼칠 권리는 없다' 라고 규정되어있습니다.

반려견을 키우는 국민은 동물세를 내며, 이 세금으로 동물 보호 및 복지 혜택 등이 제공됩니다. 반려견의 품종과 지역에 따라 연간 약 24~100유로의 세금이 부과됩니다. 반려동물의 매매 는 법적으로 금지되어있어 일반적으로 인증기관(브리더 협회) 의 검증을 받은 보호자 가정이나 독일의 최대 규모의 유기 동물 보호소인 '티어하임'을 통해 입양할 수 있습니다. 티어하임은 모든 축사에 자연광이 들어오고 엄격한 청결 관리가 이뤄집니다. 안락사 자체가 불법이기 때문에 설사 입양이 되지 않더라도 불치병과 같은 큰 고통이 지속 되지 않는 이상 자연사 할 때까지 보살핍니다.

반려견을 입양한 견주는 강아지 학교 '훈데슐레'에 다니면서 반려견과 사람이 함께 지내는 법에 대한 교육을 받습니다.

According to German animal protection law Chapter 1 Article 1, 'Animals are the same creatures as human. No one has the right to harm an animal without a justifiable reason.'

A citizen who owns a companion dog pays animal tax and through this tax, is provided animal protection, welfare benefits etc. Depending on the breed and the region, charged tax may vary from 24 to 100€ a year. Buying or selling a companion animal is illegal and companion animal can be adopted through an authorized family (authorized by breeder association) or through Tierheim, the biggest animal shelter in Germany. Every kennel in Tierheim has a window to bring in sunlight and is kept clean with strict sanitary rules. As euthanization is legally prohibited, the animals are looked after till natural cause of death in the facility, even if adoption doesn't carry out, as long as they don't suffer from constant pain such as cancer.

A person who has adopted a dog must go to Hundeschule, a dog school, to learn how to live along with them.

L'Hoest's Monkey,
a huge beard
거대한 턱수염, 로에스트 원숭이

Size 크기	45~70cm / 3~10kg	**Lifespan** 평균수명	30 years in captivity 사육시 최대 30년
Diet 식성	Herbivore 초식	**Habitat** 서식지	Forest 숲
Behavior 행동	Group living 무리생활	**Distribution** 분포지역	Congo, Rwanda, Uganda 콩고, 르완다, 우간다
Reproduction 번식	Average 1 평균 1마리	**Conservation status** 멸종위기등급	Vulnerable (VU) 취약종

Lion,
kings of beasts
동물의 왕, 사자

Size 크기	240~330cm / 126~272kg	**Lifespan** 평균수명	14 years
Diet 식성	Carnivore 육식	**Habitat** 서식지	Savana, grassland 사바나, 초원
Behavior 행동	Group living 무리생활	**Distribution** 분포지역	Sub-saharan Africa 사하라 이남 아프리카
Reproduction 번식	Average 3 평균 3마리	**Conservation status** 멸종위기 등급	Vulnerable (VU) 취약종

Meerkat,
the watchman of desert
사막의 파수꾼, 미어캣

Size 크기	42.5~60cm / 0.7kg
Diet 식성	Omnivore 잡식
Behavior 행동	Group living 무리 생활
Reproduction 번식	Average 4 평균 4마리
Lifespan 평균수명	10 years
Habitat 서식지	Desert, dune 사막, 언덕
Distribution 분포지역	Botswana, Zimbabwe Mozambique, South Africa 보츠나와, 짐바브웨, 모잠비크, 남아프리카
Conservation status 멸종위기 등급	Least concern (LC) 관심대상종

MEERKAT
LEAST CONCERN SPECIES

- EX Extinct
- EW Extinct in the Wild
- CR Critically Endangered
- EN Endangered
- VU Vulnerable
- NT Near Threatened
- LC Least Concern

Ibex,
fighter on the cliff
절벽 위의 싸움꾼, 아이벡스

Size 크기	130~140m / 65~100kg	**Life span in wild** 평균수명	15 years
Diet 식성	Herbivore 초식	**Habitat** 서식지	Scrub forest, mountain 관목산림, 산
Behavior 행동	Group living 무리생활	**Distribution** 분포지역	China, Ethiopia, Europe 중국, 에티오피아, 유럽
Reproduction 번식	Average 1 평균 1마리	**Conservation status** 멸종위기 등급	Vulnerable (VU) 취약종

A male ibex fights with other males to win females in the herd.
수컷은 무리 내 암컷을 차지하기 위해 다른 수컷과 싸웁니다.

Gorilla,
bodybuilder of rainforest
열대우림의 보디빌더, 고릴라

Size 크기	130~180cm / 275kg
Diet 식성	Herbivore 초식
Behavior 행동	Group living 무리생활
Reproduction 번식	Average 1 평균 1마리
Lifespan 평균수명	40~50 years
Habitat 서식지	Rainforest, forest 열대우림, 숲
Distribution 분포지역	Central African Republic Gabon, Nigeria, Angola 중앙아프리카공화국, 가봉, 나이지리아, 앙골라
Conservation status 멸종위기 등급	Critically endangered (CR) 위급종

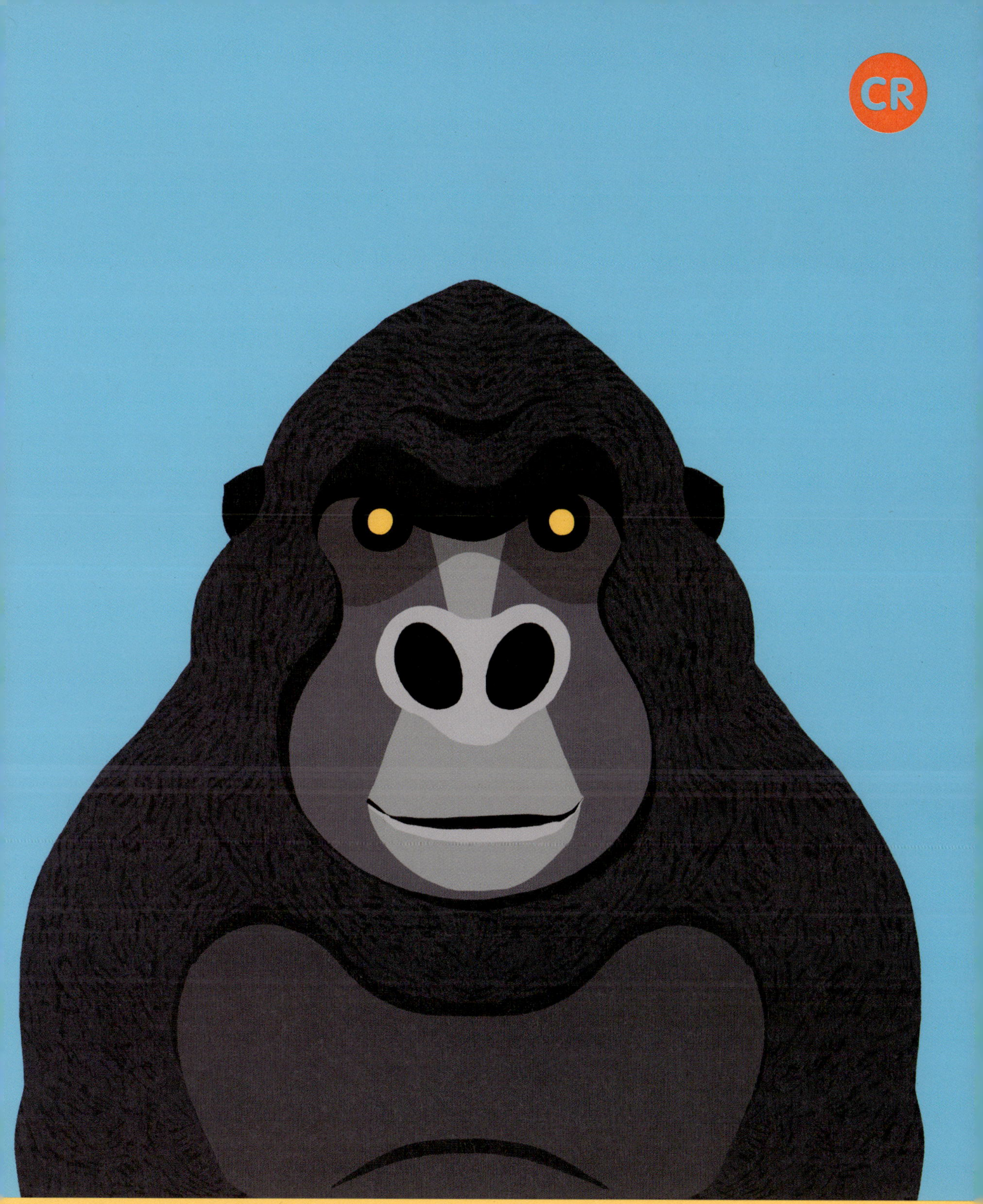

EX Extinct
EW Extinct in the Wild
CR Critically Endangered
EN Endangered
VU Vulnerable
NT Near Threatened
LC Least Concern

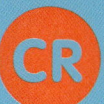

GORILLA
CRITICALLY ENDANGERED SPECIES

Otter,
the architect of wildlife
야생의 건축가, 수달

Size 크기	63~75cm / 5.8~10kg	**Lifespan** 평균수명	22 years in captivity 사육시 최대 22년
Diet 식성	Fish, Crustaceans, Clam 어류, 갑각류, 조개	**Habitat** 서식지	River, riverside 강, 물가
Behavior 행동	Solitary living 단독생활	**Distribution** 분포지역	Eurasia, North Africa 유라시아, 북 아프리카
Reproduction 번식	Average 2 평균 2마리	**Conservation status** 멸종위기 등급	Near threatened (NT) 준위협종

OTTER
NEAR THREATENED SPECIES

- EX Extinct
- EW Extinct in the Wild
- CR Critically Endangered
- EN Endangered
- VU Vulnerable
- NT Near Threatened
- LC Least Concern

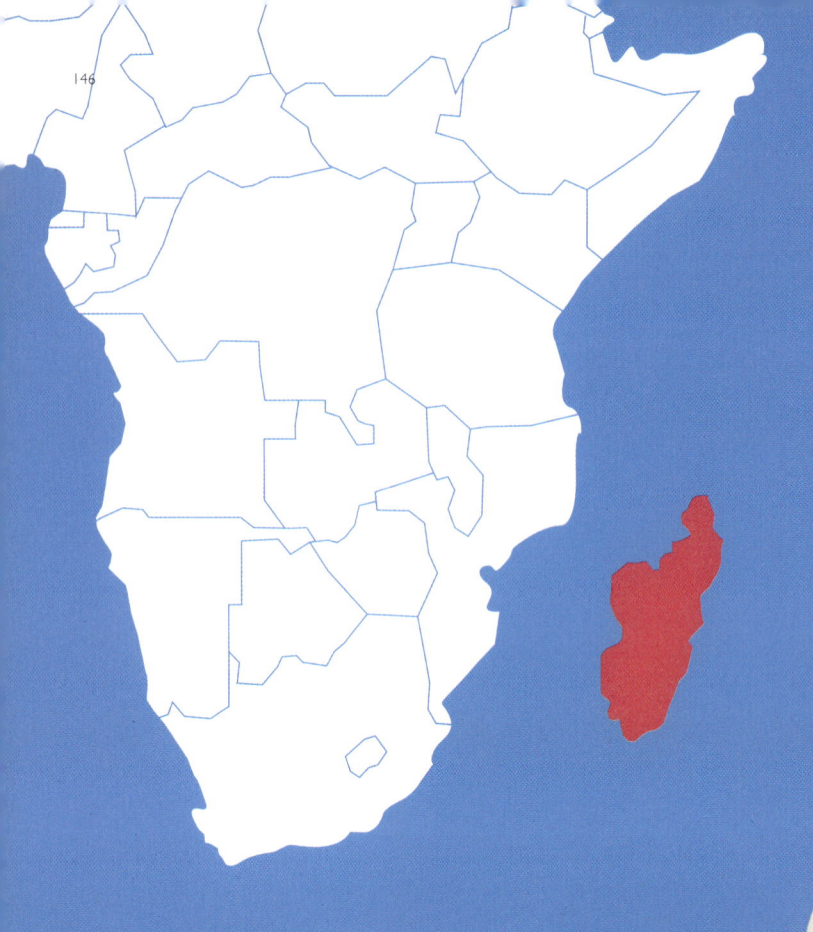

Coquerel's sifaka,
the dancing queen of jungle
정글의 댄싱퀸, 코쿠렐시파카

Size 크기	30~50cm / 3.7~4.3kg	**Lifespan** 평균수명	27 years
Diet 식성	Herbivore 초식	**Habitat** 서식지	Rainforest, forest 열대우림, 숲
Behavior 행동	Group living 무리생활	**Distribution** 분포지역	Madagascar 마다가스카르
Reproduction 번식	Average 1 평균 1마리	**Conservation status** 멸종위기 등급	Endangered (EN) 위기종

When it walks, a Coquerel's sifaka scapers as if it were dancing.
춤을 추듯 깡충거리며 걷습니다.

COQUEREL'S SIFAKA
CRITICALLY ENDANGERED SPECIES

EX Extinct
EW Extinct in the Wild
CR Critically Endangered
EN Endangered
VU Vulnerable
NT Near Threatened
LC Least Concern

ANIMAL PROTECTION LAWS BY COUNTRY : SWITZERLAND

국가별
동물보호법 :
스위스

스위스는 토끼 사육장의 크기, 개에게 제공해야 하는 운동량 등을 포함한 관리 규제가 세계에서 가장 엄격합니다. 1992년 헌법을 개정하면서 동물을 사물이 아닌 생명으로 규정되어 반려 동물을 처음 기르는 반려인들은 관련 교육을 필수로 이수해야하며 동물보호법 위반 시 최대 3년 이하의 징역 또는 20,000 스위스프랑의 벌금이 부과됩니다.

Switzerland has the strictest regulations in the world that even include the size of a rabbit's cage and the necessary amount of exercise for dogs. When the constitution was revised in 1992, because animals were defined as living creatures, not objects, all animal owners must finish a course in how to raise a companion animal. Also, in case of violating animal protection law one can be sentenced up to 3 years in prison or fined 20,000 Swiss francs.

ANIMAL PROTECTION LAWS BY COUNTRY : U.S.A.

국가별 동물보호법 : 미국

미국의 경우 반려 동물의 역사가 100년 가까이 되며 주마다 다소 차이가 있으나 대다수 주에서 동물 등록제를 실시하고 있습니다. 또한 위험한 개에 대한 기준, 공공장소에서의 기준 등 각 주마다 동물에 대한 조례가 만들어져 있습니다.

미국 내 동물의 운송을 담당하는 업자들은 이동중 동물에게 충분한 음식과 물을 제공해야하고 휴식없이 연속으로 28시간 이상 가두지 못하도록 규정하였습니다. 군용 동물로 수명이 다하거나 기능을 하지 못하는 경우에는 입양할 수 있도록 허가하는 '군용 동물 입양법' 등이 있습니다.

United States have a history of companion animal for almost 100 years. Although they have a bit of differences depending on the state, but most of the states are practicing animal registration policy. Also, every state has a regulation about animals such as, standard of a dangerous dog, public places and so on.

Animals that are transported within the US territories must be provided enough food and water and cannot be locked up for 28 consecutive hours without rest. There is also a 'military animal adoption legislation' that allows them to be adopted if they cannot be used as military use any more.

De brazza's monkey,
French explorer

프랑스 탐험가, 드 브라자

Size 크기	40~63.5cm / 4~7kg
Diet 식성	Omnivore 잡식
Behavior 행동	Group living 무리생활
Reproduction 번식	Average 1 평균 1마리
Lifespan 평균수명	30 years in captivity 사육시 평균 30년
Habitat 서식지	Rainforest 열대우림
Distribution 분포지역	Angola, Cameroon Equatorial Guinea 앙골라, 카메룬, 적도 기니
Conservation status 멸종위기 등급	Least concern (LC) 관심대상종

It was named after the French explorer who first found a Brazzaville in the Republic of the Congo.

콩고공화국의 수도인 브라자빌을 처음 발견한 프랑스 탐험가 '브라자'의 이름을 따 지어졌습니다.

Sand cat,
the Peter Pan of desert
사막의 피터팬, 모래고양이

Size 크기	45~57cm / 1.4~3.4kg	**Lifespan** 평균수명	13 years in captivity 사육시 평균 13년
Diet 식성	Carnivore 육식	**Habitat** 서식지	Desert, dune 사막, 언덕
Behavior 행동	Solitary living 단독생활	**Distribution** 분포지역	Central Asia, Algeria, Niger, Morocco 중앙아시아, 알제리, 니제르, 모로코
Reproduction 번식	Average 4 평균 4마리	**Conservation status** 멸종위기 등급	Least concern (LC) 관심대상종

The appearance of a sand cat does not change as it gets old.
나이가 들어도 외모가 변하지 않습니다.

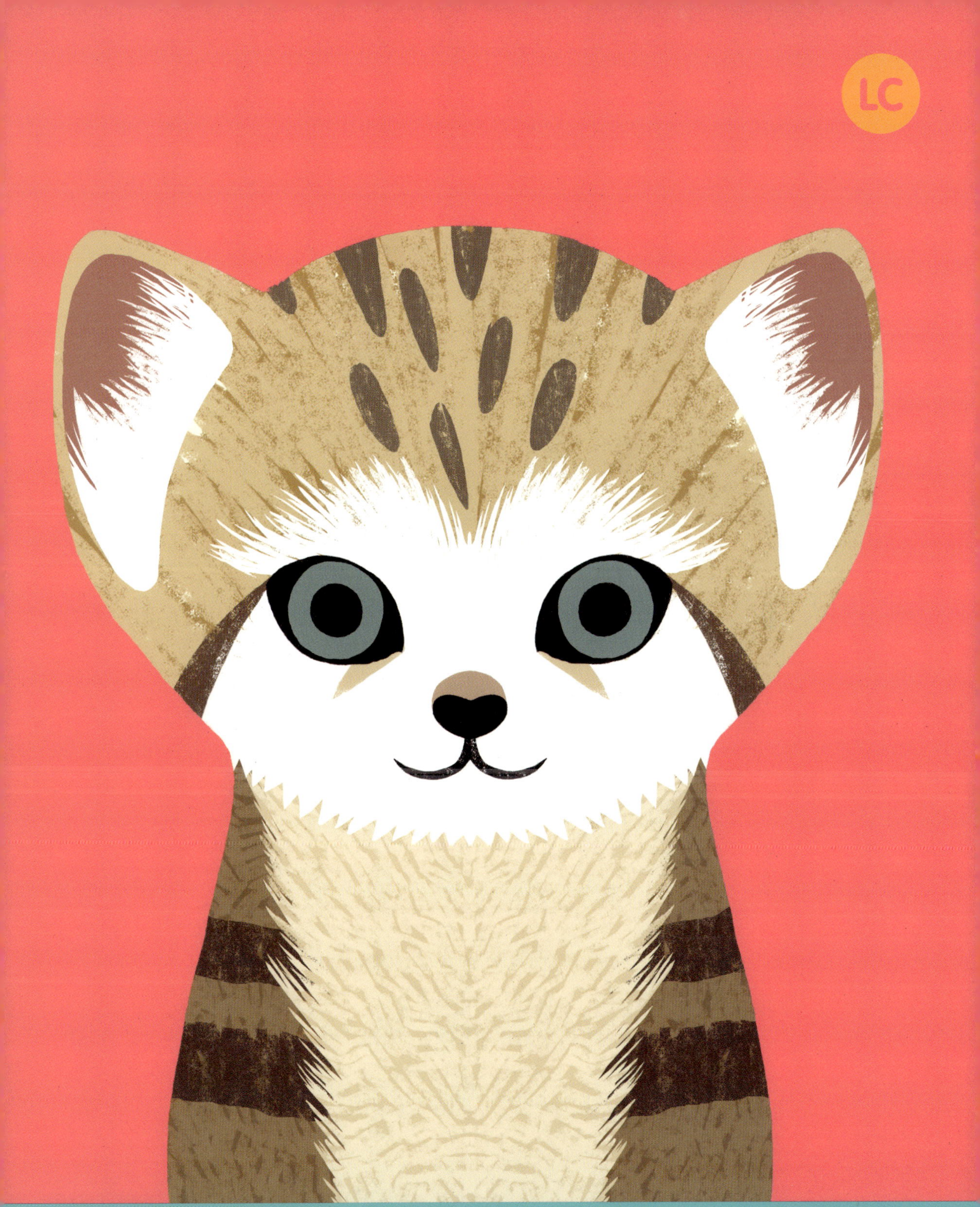

SAND CAT
LEAST CONCERN SPECIES

- EX Extinct
- EW Extinct in the Wild
- CR Critically Endangered
- EN Endangered
- VU Vulnerable
- NT Near Threatened
- LC Least Concern

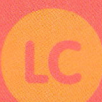

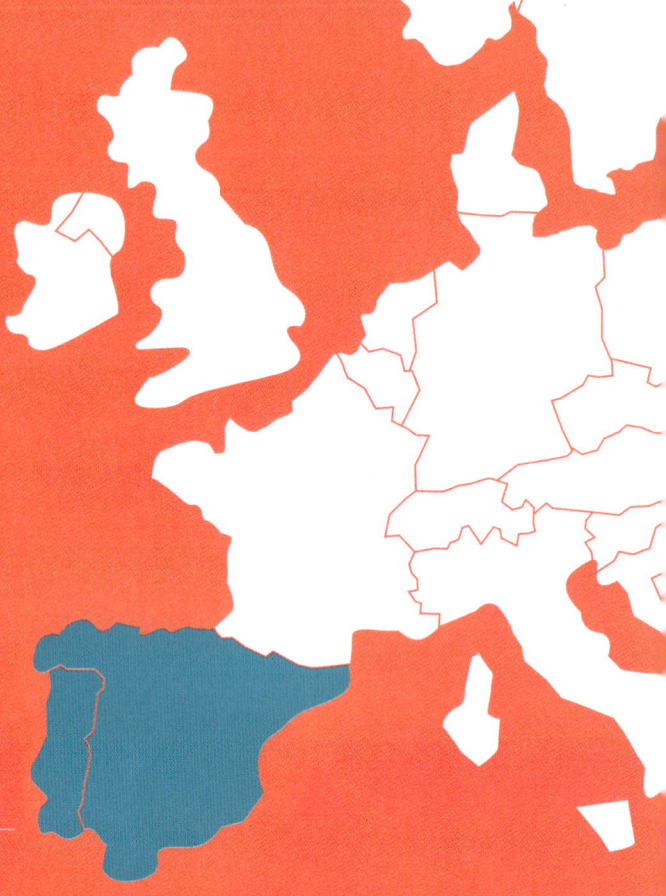

Iberian lynx,
charismatic pointy ears

카리스마 뾰족귀, 이베리아 스라소니

Size 크기	75~82cm / 7~16kg
Diet 식성	Carnivore 육식
Behavior 행동	Solitary living 단독생활
Reproduction 번식	Average 2~3 평균 2~3마리
Lifespan 평균수명	13 years
Habitat 서식지	Savana, grassland 사바나, 초원
Distribution 분포지역	Iberian Peninsula in southwestern Europe 남서유럽 이베리아 반도
Conservation status 멸종위기등급	Endangered (EN) 위기종

Aye-aye, nocturnal monkey
밤의 원숭이, 아이아이 원숭이

Size 크기	30~40cm / 2~2.7kg	**Lifespan** 평균수명	10~20 years
Diet 식성	Omnivore 잡식	**Habitat** 서식지	Tropical forest 열대우림
Behavior 행동	Solitary living 단독생활	**Distribution** 분포지역	Madagascar 마다가스카르
Reproduction 번식	Average 1 평균 1마리	**Conservation status** 멸종위기등급	Endangered (EN) 위기종

Although it has a frightening outer feature, it's actually a docile animal that feeds on bamboo shoots and bird eggs.
아이아이 원숭이는 희귀한 생김새로 사람들에게 두려움을 샀지만 죽순과 조류의 알을 먹는 순한 동물입니다.

AYE-AYE
ENDANGERED SPECIES

EX Extinct
EW Extinct in the Wild
CR Critically Endangered
EN Endangered
VU Vulnerable
NT Near Threatened
LC Least Concern

Indri,
the biggest lemuriformes
가장 큰 여우원숭이, 인드리 원숭이

Size 크기	64~72cm / 6~9.5kg	**Lifespan** 평균수명	15~22 years
Diet 식성	Herbivore 초식	**Habitat** 서식지	Lowland rainforest 저지대 열대우림
Behavior 행동	Group living 무리생활	**Distribution** 분포지역	Eastern Madagascar 마다가스카르 동부
Reproduction 번식	Average 1 평균 1마리	**Conservation status** 멸종위기등급	Critically endangered (CR) 위급종

African golden cat,
leopards's bro
표범의 형제, 아프리카 황금고양이

Size 크기	61~110cm / 4~18kg	**Lifespan** 평균수명	12 years
Diet 식성	Carnivore 육식	**Habitat** 서식지	Tropical forest 열대우림
Behavior 행동	Solitary living 단독생활	**Distribution** 분포지역	Equatorial Africa 적도 아프리카
Reproduction 번식	Average 2 평균 2마리	**Conservation status** 멸종위기등급	Vulnerable (VU) 취약종

Because African golden cat shares habitats with leopards, they are called "leopard's bro".
아프리카 황금고양이는 표범과 서로의 서식지를 공유하기 때문에 "표범의 황제"라고 불립니다.

AFRICAN GOLDEN CAT
VULNERABLE SPECIES

- EX Extinct
- EW Extinct in the Wild
- CR Critically Endangered
- EN Endangered
- VU Vulnerable
- NT Near Threatened
- LC Least Concern

Red-tailed Monkey,
a heart-shaped nose full of love

사랑이 가득한 하트 코, 붉은꼬리원숭이

Size 크기	38~46cm / 1.4~2.9kg	**Lifespan** 평균수명	30 years
Diet 식성	Herbivore 초식	**Habitat** 서식지	Forest 숲
Behavior 행동	Group living 무리생활	**Distribution** 분포지역	Angola, Congo, Kenya 앙골라, 콩고, 케냐
Reproduction 번식	Average 1 평균 1마리	**Conservation status** 멸종위기등급	Least concern (LC) 관심대상종

Male and female red-tailed monkeys express affection through nose-to-nose contact and touching each other's noses.
암컷과 수컷은 서로의 코를 만지거나 맞대는 등 코인사로 애정표현을 합니다.

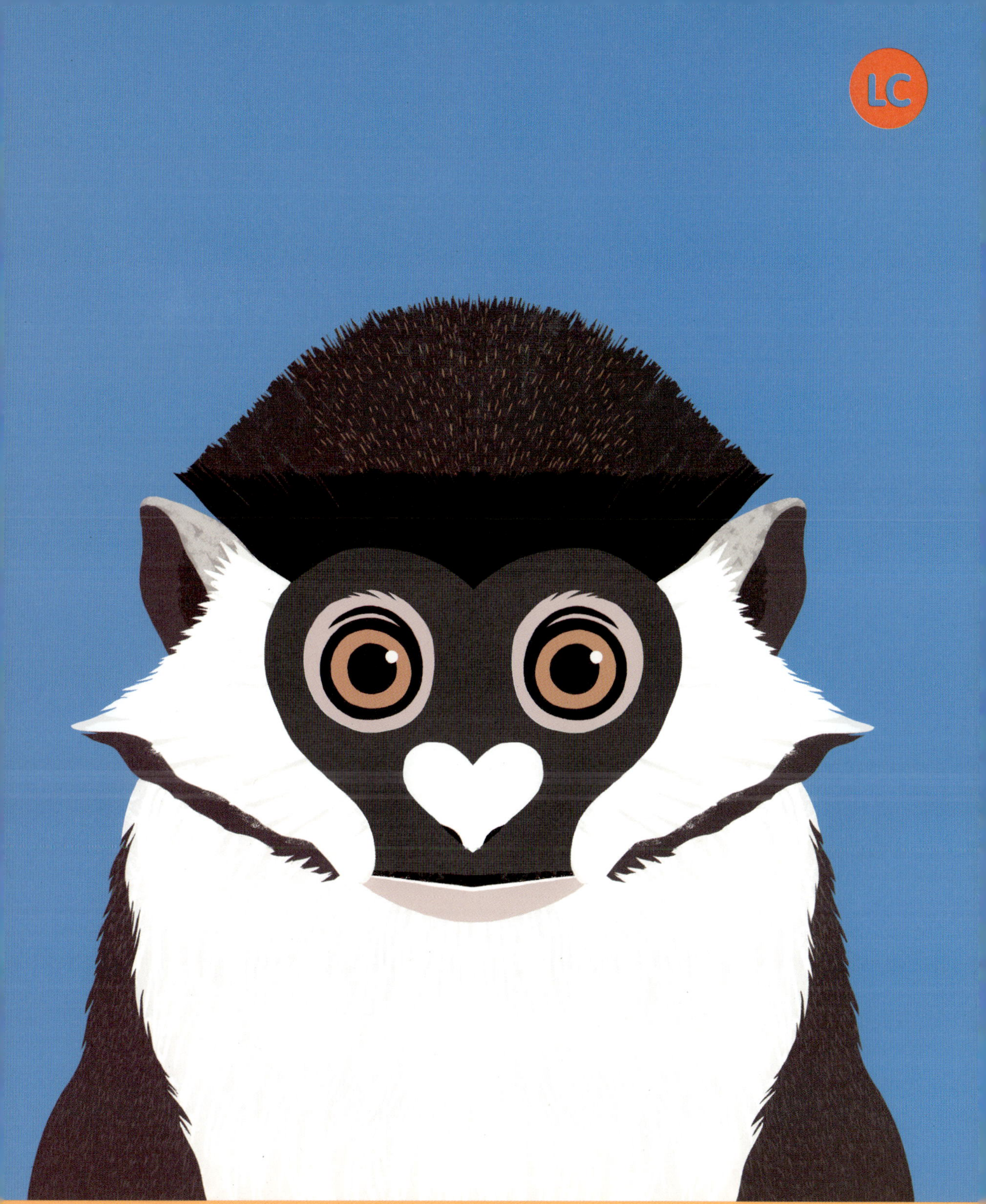

EX Extinct
EW Extinct in the Wild
CR Critically Endangered
EN Endangered
VU Vulnerable
NT Near Threatened
LC Least Concern

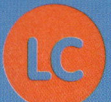

RED-TAILED MONKEY
LEAST CONCERN SPECIES

ANIMAL PROTECTION LAWS BY COUNTRY : TAIWAN

국가별 동물보호법 : 대만

아시아 국가 중 최초로 식용을 위해 개, 고양이 도살을 금지하는 법안을 제정하였습니다. 이를 어길 시 5만 ~25만 대만달러의 벌금이 부과되며 가해자의 이름과 사진, 범죄 사실이 공개 될 수 있습니다. (제 6장 27조) 또한, 고의로 동물에게 상해를 입히거나 도축이 금지된 개, 고양이 또는 동물을 죽일 경우 20만 ~ 200만 대만달러 이하의 벌금이 부과됩니다. (제6장 25조)

동물 전시 업체는 상업적 활동에 앞서 당국으로부터 면허를 취득해야 하며(제 6조 1항)며 주인은 반려동물의 출생정보 등을 관할 당국 또는 위임된 민간 단체에 등록하고 지방 관할 당국은 등록된 반려동물에게 ID태그를 발행해야 합니다. (제 4장 19조)

Taiwan is the first in Asia to legislate the laws that ban killing dogs and cats for consumption. In case of violation, one can be fined from NT$50,000 to NT$ 250,000 and the violator's name, photo, and his crime can be open to public. (chapter6 article27) If you intentionally harass a companion animal or kill a dog, cat or any other animals that have been prohibited to kill, you will be fined from NT$20,000 to NT$200,000 (chapter6 article25).

Animal displaying enterprises must be licensed by the government prior to any commercial acts (chapter6 article1) and the owner of the companion animal must register the animal's birth and other information to the authorities or to the delegated private organizations. Regional government must issue the registered animals ID tags. (Chapter4 article19)

ANIMAL PROTECTION LAWS BY COUNTRY : SINGAPORE

국가별 동물보호법 : 싱가포르

싱가포르의 경우 거주지에 따라 키울 수 있는 반려동물의 수와 종류가 제한되어있습니다. 아파트에서는 1마리, 개인 부지가 있을 경우에는 3마리까지 키울 수 있습니다.

동물 학대에 대한 유죄 판결을받은 사람들은 현재 최대 1만 싱가포르 달러의 벌금, 최대 1년 이하의 징역 또는 두 가지 모두에 처해질 수 있습니다.

In case of Singapore, the number and the kind of companion animal available is restricted depending on one's house. In an apartment, only 1 animal is allowed, and in case of owning separate site, 3 are allowed.

Today, anyone who was found guilty of animal harassment can be fined up to 10,000 SGD, sentenced 1 year in prison or both.

Crowned lemur,
orange crown
오렌지 왕관, 관여우원숭이

Size 크기	31~36cm / 2~4kg	**Lifespan** 평균수명	20~36 years
Diet 식성	Herbivore 초식	**Habitat** 서식지	Dry decidous forest 건조지역의 낙엽성 숲
Behavior 행동	Group living 무리생활	**Distribution** 분포지역	Northern tip of Madagascar 마다가스카르 북쪽 끝
Reproduction 번식	Average 1~2 평균 1~2마리	**Conservation status** 멸종위기등급	Endangered (EN) 위기종

There's a clear difference between female Crowned lemur and male Crowned lemur. While the female has a silvery grey colored body, male has a dark red brown body. 관여우원숭이는 암컷과 수컷이 확연한 생김새의 차이를 보입니다. 암컷은 은빛 회색의 몸을, 수컷은 어두운 빛깔의 붉은 갈색 몸을 가지고있습니다.

CROWNED LEMUR
ENDANGERED SPECIES

EX Extinct
EW Extinct in the Wild
CR Critically Endangered
EN Endangered
VU Vulnerable
NT Near Threatened
LC Least Concern

Addax,
hornderful

아름다운 뿔, 아닥스

Size 크기	150~170cm / 60~125kg	**Lifespan** 평균수명	25 years in captivity 사육시 최대 25년
Diet 식성	Herbivore 초식	**Habitat** 서식지	Savana, grassland 사바나, 초원
Behavior 행동	Group living 무리생활	**Distribution** 분포지역	Morocco, Tunisia 모로코, 튀니지
Reproduction 번식	Average 1 평균 1마리	**Conservation status** 멸종위기등급	Critically endangered (CR) 위급종

Ethiopian wolf,
the only wolf in Africa
아프리카 유일 늑대, 에티오피아 늑대

Size 크기	90~110cm / 10~20kg	**Lifespan** 평균수명	10~12 years
Diet 식성	Carnivore 육식	**Habitat** 서식지	Highland 산악지대
Behavior 행동	Solitary living 단독생활	**Distribution** 분포지역	Ethiopia 에티오피아
Reproduction 번식	Average 6 평균 6마리	**Conservation status** 멸종위기등급	Endangered (EN) 위기종

ETHIOPIAN WOLF
ENDANGERED SPECIES

EX Extinct
EW Extinct in the Wild
CR Critically Endangered
EN Endangered
VU Vulnerable
NT Near Threatened
LC Least Concern

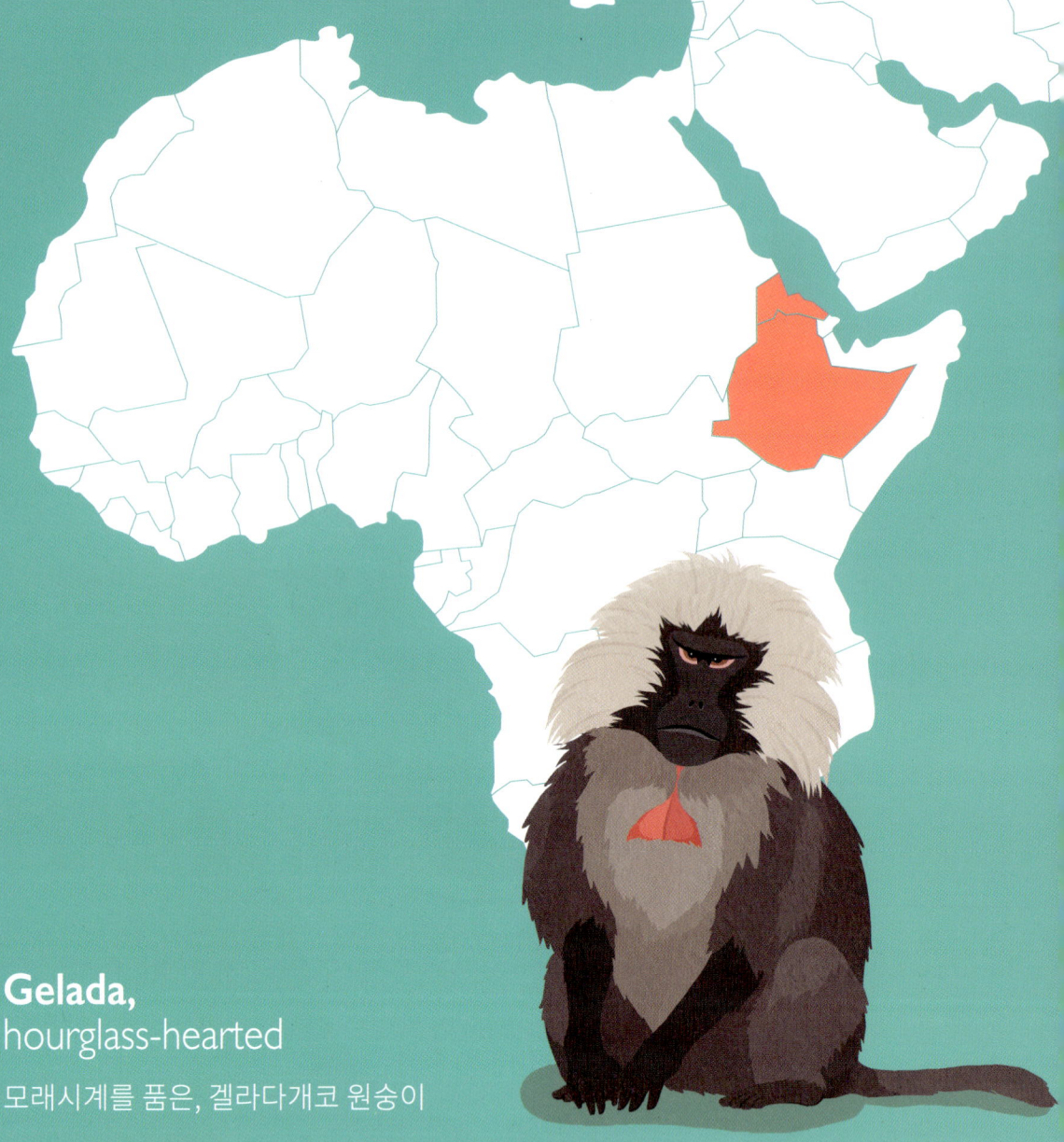

Gelada,
hourglass-hearted

모래시계를 품은, 겔라다개코 원숭이

Size 크기	50~74cm / 13~21kg	**Lifespan** 평균수명	20~30 years
Diet 식성	Omnivore 잡식	**Habitat** 서식지	Savana 사바나
Behavior 행동	Group living 무리생활	**Distribution** 분포지역	Ethiopia, Eritrea 에티오피아, 에리트레아
Reproduction 번식	Average 1 평균 1마리	**Conservation status** 멸종위기 등급	Least concern (LC) 관심대상종

Both the Gelada female and male have hourglass-shaped patch on their chests.
겔라다개코원숭이는 암컷과 수컷 모두 가슴에 모래시계 모양의 패치를 지녔습니다.

Wolf's mona monkey,
a wolf not related to wolf
늑대와는 관련없는, 늑대 모나 원숭이

Size 크기	44~51cm / 2.5~4.5kg	Lifespan 평균수명	20~26 years
Diet 식성	Omnivore 잡식	Habitat 서식지	Forest 숲
Behavior 행동	Group living 무리생활	Distribution 분포지역	Central Africa 중앙아프리카카
Reproduction 번식	Average 1 평균 1마리	Conservation status 멸종위기 등급	Near threatened (NT) 준위협종

The name 'Wolf's monkey' was taken from Dr. Ludwig Wolf, who first discovered the species, and has no biological relation to wolves.
최초 발견한 Ludwig Wolf 박사의 이름을 붙여 "Wolf's monkey"라 지어졌으며, 늑대와는 아무런 생물학적 관련이 없습니다.

WOLF'S MONA MONKEY
NEAR THREATENED SPECIES

- EX Extinct
- EW Extinct in the Wild
- CR Critically Endangered
- EN Endangered
- VU Vulnerable
- NT Near Threatened
- LC Least Concern

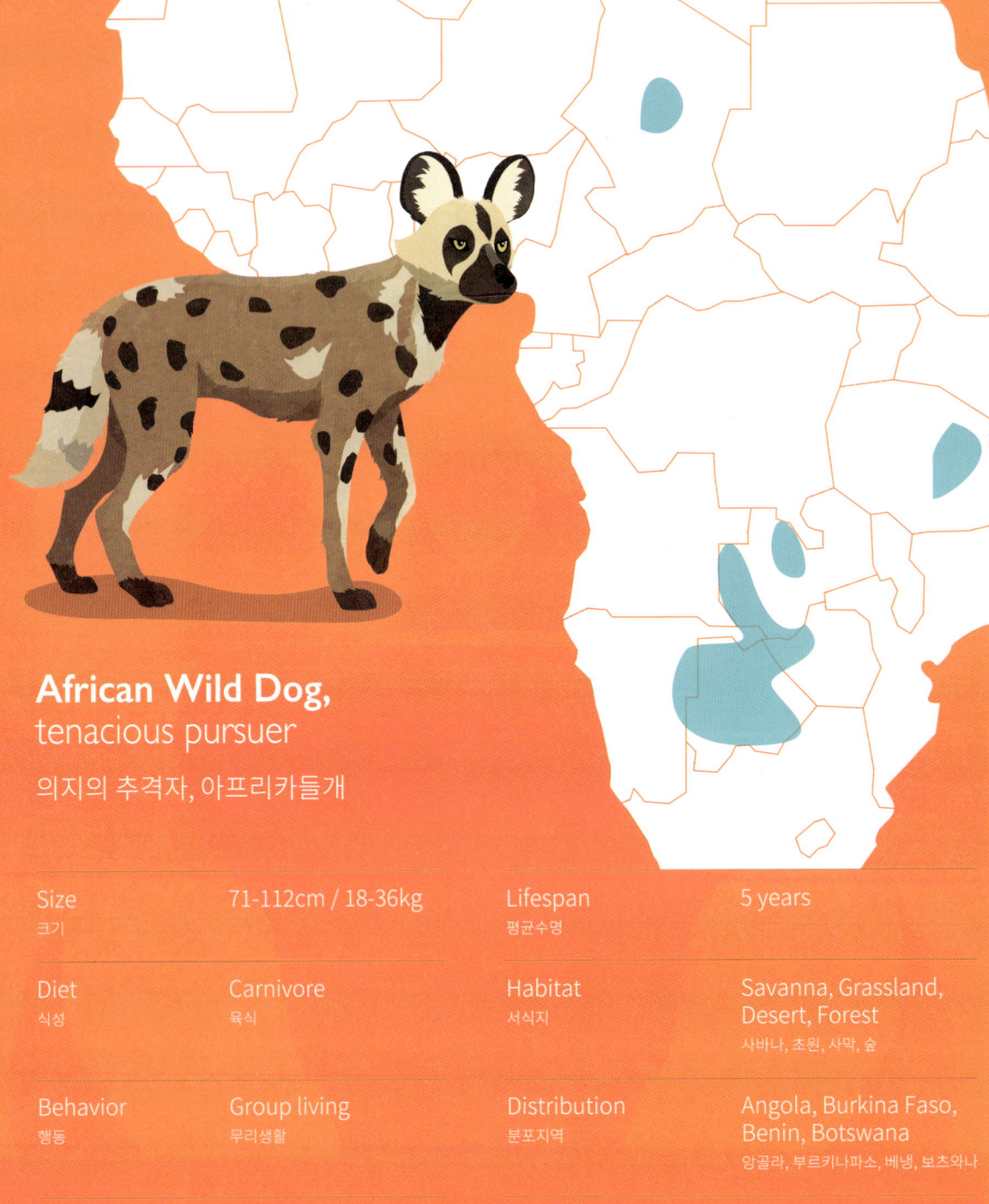

African Wild Dog,
tenacious pursuer
의지의 추격자, 아프리카들개

Size 크기	71-112cm / 18-36kg	**Lifespan** 평균수명	5 years
Diet 식성	Carnivore 육식	**Habitat** 서식지	Savanna, Grassland, Desert, Forest 사바나, 초원, 사막, 숲
Behavior 행동	Group living 무리생활	**Distribution** 분포지역	Angola, Burkina Faso, Benin, Botswana 앙골라, 부르키나파소, 베냉, 보츠와나
Reproduction 번식	Average 2~8 평균 2~8마리	**Conservation status** 멸종위기 등급	Endangered (EN) 위기종

African wild dogs are experts in chasing and hunting their prey until it becomes exhausted and dehydrated.
아프리카들개는 먹잇감이 탈진할 때까지 추격해 사냥하는 데 전문가입니다.

AFRICAN WILD DOG
ENDANGERED SPECIES

EX Extinct
EW Extinct in the Wild
CR Critically Endangered
EN Endangered
VU Vulnerable
NT Near Threatened
LC Least Concern

Ring tailed lemur,
spick-and-span monkey
깔끔쟁이, 알락꼬리 여우 원숭이

Size 크기	38~45cm / 2.3~3.5kg	**Lifespan** 평균수명	27 years
Diet 식성	Omnivore 잡식	**Habitat** 서식지	Forest, rainforest 숲, 열대우림
Behavior 행동	Group living 무리생활	**Distribution** 분포지역	Southwestern Madagascar 마다가스카르 남서쪽
Reproduction 번식	Average 1 평균 1마리	**Conservation status** 멸종위기 등급	Endangered (EN) 위기종

To avoid juice from failling on its furry body a ring tailed lemur hardly needs
to use its hands, it can just raise its head while eating.
식물의 즙이 떨어지는 것을 피하기 위해 머리를 높이 들고 먹고 먹이를 쥐거나 잡는데 거의 자신의 손을 사용하지 않습니다.

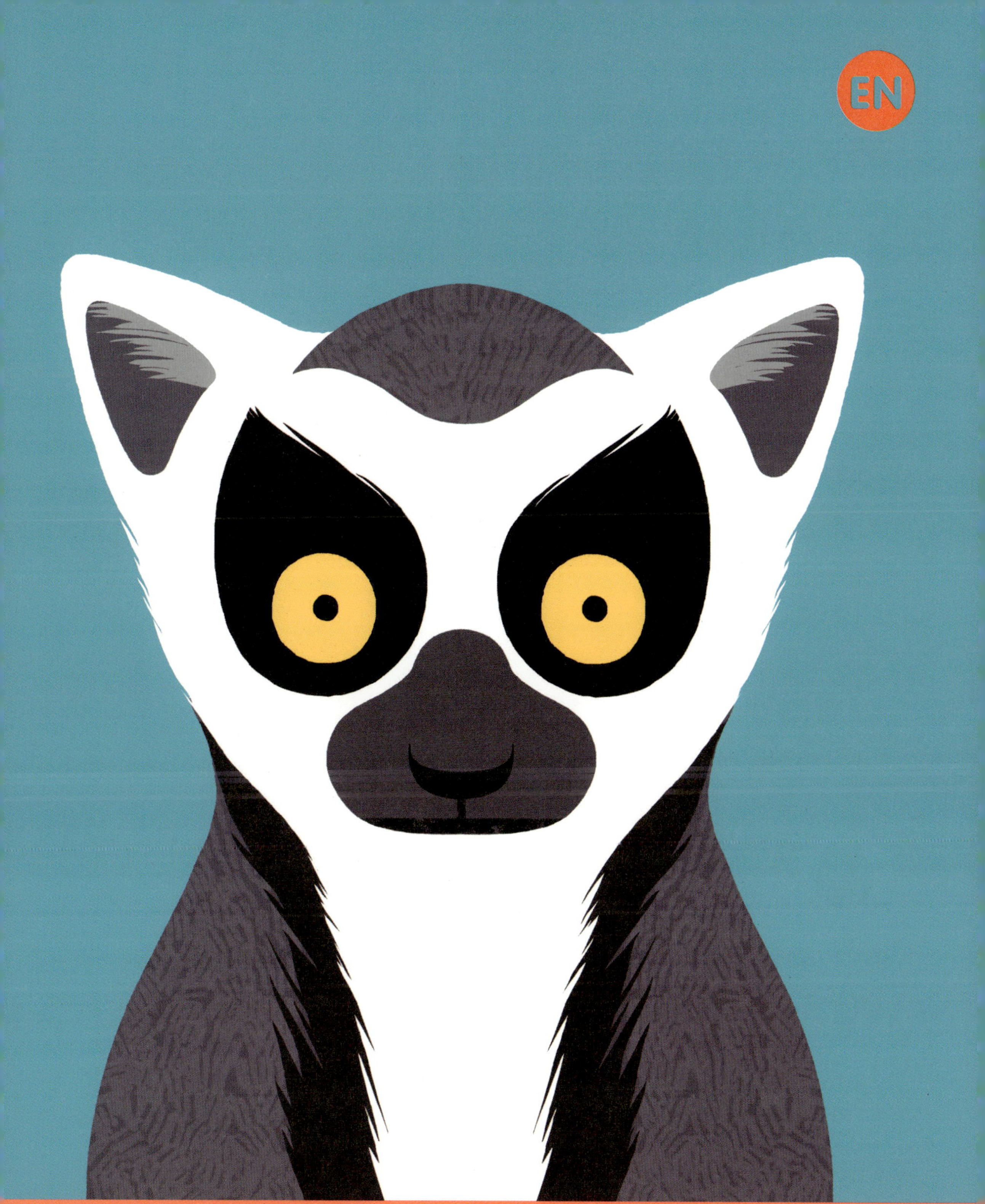

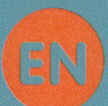

RING TAILED LEMUR
ENDANGERED SPECIES

- EX Extinct
- EW Extinct in the Wild
- CR Critically Endangered
- EN **Endangered**
- VU Vulnerable
- NT Near Threatened
- LC Least Concern

ANIMAL PROTECTION LAWS BY COUNTRY : JAPAN

국가별 동물보호법 : 일본

1973년 동물보호법이 제정되었으며, "동물을 해치거나 동물 학대를 자행해서는 안된다."라고 명시되어 있습니다. 특히 모든 포유 동물에 대한 학대를 범죄로 규정합니다. 정당한 사유없이 동물을 사망 시키거나 상해를 입히면 1년 이하의 징역 또는 최대 1백만엔 이하의 벌금이 부과됩니다. 반려동물에게 먹이를 주지 않아 병들게 할 경우, 또는 그러한 형태의 학대를 할 경우 최대 50만엔의 과태료가 부과됩니다.

2006년 제정된 동물 애호 치료 및 관리법은 동물 관련 실험을 규제합니다. 이 법은 '3R' 원칙을 준수해야 함을 강조합니다. 3R이란 실험동물 이외의 방법으로 대체(Replacement), 실험동물 숫자 감축 (Reduction), 실험동물의 고통 경감(Refinement)을 의미합니다.

In 1973, animal protection law has been passed and it is stated that "no one can harm or harass an animal". Japan especially defines harassing a mammal as a crime. If one is to kill or harm an animal without a justifia -ble cause, one can be sentenced to a year in prison or fined to 1 million yens. If one does not feed the companion animal, make it ill, or carry out any harassment in suc -h ways, one can be fined to 500,000 yens.

Animal protection treatment and management law which has been passed in 2006 limits animal experiments. This law emphasizes that experimenters must follow the 3R principle. 3R means one must look for other methods to "replace" the animal experiment, "reduce" the number of animals used, and "refine" the animal's pain.

ANIMAL PROTECTION LAWS BY COUNTRY : SOUTH KOREA

국가별 동물보호법 : 대한민국

대한민국의 동물보호법은 다른 국가들에 비해 법적으로 미약한 부분이 있었으나 여러 동물보호단체와 시민들의 노력으로 2022년 4월, 31년만에 큰 개정안이 발표되었습니다. 동물학대 행위자에 대한 처벌은 3년 이하 징역 또는 3천만원 이하의 벌금으로 강화되었으며 민간 동물보호 시설 신고제가 도입되었습니다.

반려 동물 영업 준수사항을 어기는 경우에는 행정 처분이 아닌 형사처벌로 변경되었으며 무허가 동물생산업·수입업·판매업·장묘업 영업 행위에 대한 처벌은 기존 500만원 벌금형에서 2년 징역, 2천만원 벌금형으로 상향되었습니다. 또한 동물 실험 시행기관에는 실험 동물의 건강을 점검하는 전임 수의사를 배치해야하고 동물 실험 윤리위원회의 동물 실험에 대한 심의 및 지도, 감독 기능도 한층 강화되었습니다.

Although the animal protection law in South Korea had been relatively weak compared to other countries, thanks to the efforts of several animal welfare group and citizens, a major revision was announced in April 2022 after 31 years. Punishment for animal abuse has been enhanced to up to 3 years imprisonment or a fine of up to about 25,000 dollars, and a reporting system for private animal shelters has been introduced.

Penalties for non-compliance by pet-related businesses now include criminal charges instead of administrative sanctions. Punishment for unlicensed animal-related businesses has increased to 2 years imprisonment or a fine of 16,000 dollars . Animal testing institutions must now appoint a full-time veterinarian and enhance the functions of the animal testing ethics committee.

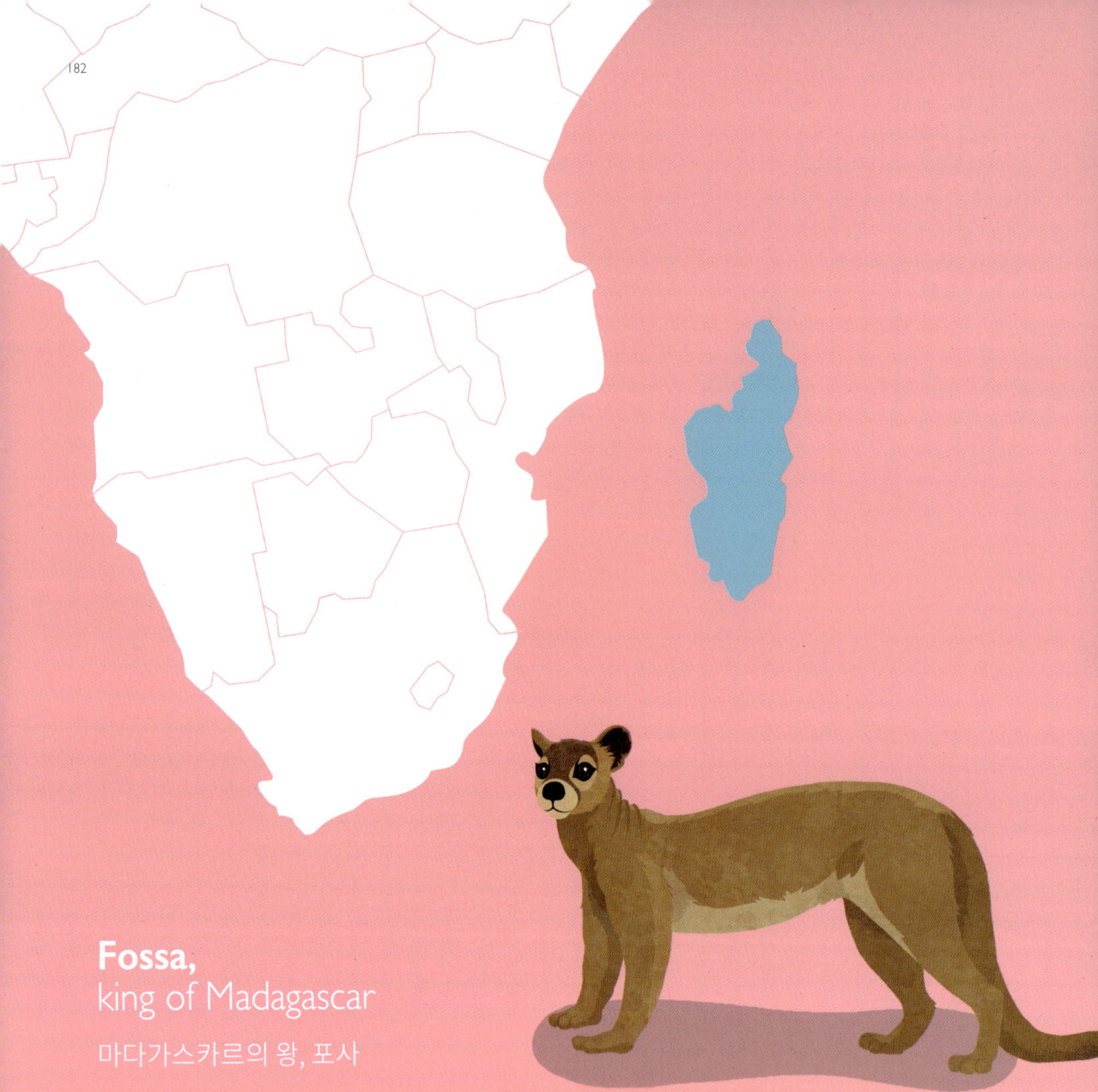

Fossa,
king of Madagascar
마다가스카르의 왕, 포사

Size 크기	60~80cm / 7~12kg	**Lifespan** 평균수명	15~20 years
Diet 식성	Carnivore 육식	**Habitat** 서식지	Tropical forest 열대우림
Behavior 행동	Solitary living 단독생활	**Distribution** 분포지역	Madagascar 마다가스카르
Reproduction 번식	Average 1 평균 1마리	**Conservation status** 멸종위기등급	Vulnerable (VU) 취약종

Fossa is the biggest mammal in Madagascar.
포사는 마다가스카르 섬에서 가장 큰 포유동물입니다.

FOSSA
VULNERABLE SPECIES

EX Extinct
EW Extinct in the Wild
CR Critically Endangered
EN Endangered
VU Vulnerable
NT Near Threatened
LC Least Concern

Red-fronted lemur,
weird hair designer
독특한 헤어 디자이너, 붉은이마 여우원숭이

Size 크기	35~48cm / 2~2.5kg	**Lifespan** 평균수명	20~30 years
Diet 식성	Omnivore 잡식	**Habitat** 서식지	Rainforest 열대우림
Behavior 행동	Group living 무리생활	**Distribution** 분포지역	Madagascar 마다가스카르
Reproduction 번식	Average 1 평균 1마리	**Conservation status** 멸종위기등급	Vulnerable (VU) 취약종

Red-fronted lemur does its hair with its bulging 6 small teeth.
붉은이마 여우원숭이는 툭 튀어나온 6개의 작은 치아로 털을 손질합니다.

RED-FRONTED LEMUR
VULNERABLE SPECIES

EX Extinct
EW Extinct in the Wild
CR Critically Endangered
EN Endangered
VU Vulnerable
NT Near Threatened
LC Least Concern

Great blue turaco, fan-head
머리 위 부채, 큰파랑 부채머리새

Size 크기	70~75cm / 0.8~1.2kg
Diet 식성	Fruit, leaves, flower 과일, 잎, 꽃
Behavior 행동	Group living 무리생활
Reproduction 번식	Average 2~3eggs 평균 2~3개의 알
Lifespan 평균수명	30 years in captivity 사육시 30년
Habitat 서식지	Tropical forest 열대우림
Distribution 분포지역	West Africa 아프리카 서쪽 지역
Conservation status 멸종위기등급	Least concern (LC) 관심대상종

Giant Dragon Lizard,
fully armored

온몸에 갑옷을 두른, 자이언트 드래곤 도마뱀

Size 크기	15~18cm / 60~100g	**Lifespan** 평균수명	15 years
Diet 식성	Insect 곤충	**Habitat** 서식지	Grassland 초원
Behavior 행동	Group living 무리생활	**Distribution** 분포지역	South Africa 남아프리카
Reproduction 번식	Average 1~2 평균 1~2마리	**Conservation status** 멸종위기등급	Vulnerable (VU) 취약종

GIANT DRAGON LIZARD
VULNERABLE SPECIES

EX Extinct
EW Extinct in the Wild
CR Critically Endangered
EN Endangered
VU Vulnerable
NT Near Threatened
LC Least Concern

Egyptian vulture,
a master of tools

도구 사용의 달인, 이집트 독수리

Size 크기	47~65cm / 1.9~2.4kg	**Lifespan** 평균수명	13~14 years
Diet 식성	Omnivore 잡식	**Habitat** 서식지	Savanna, Shrubland, Grassland 사바나, 관목 지대, 초원
Behavior 행동	Solitary living 단독생활	**Distribution** 분포지역	Southern Europe, Northern Africa, Western & Southern Asia 남유럽, 북아프리카, 서남아시아
Reproduction 번식	Average 2eggs 평균 2개의 알	**Conservation status** 멸종위기등급	Endangered (EN) 위기종

Egyptian vultures also use tools such as rocks to crack open and eat the eggs of ostriches.
이집트 독수리는 조약돌과 같은 도구를 사용하여 타조의 알을 깨고 속을 먹기도 합니다.

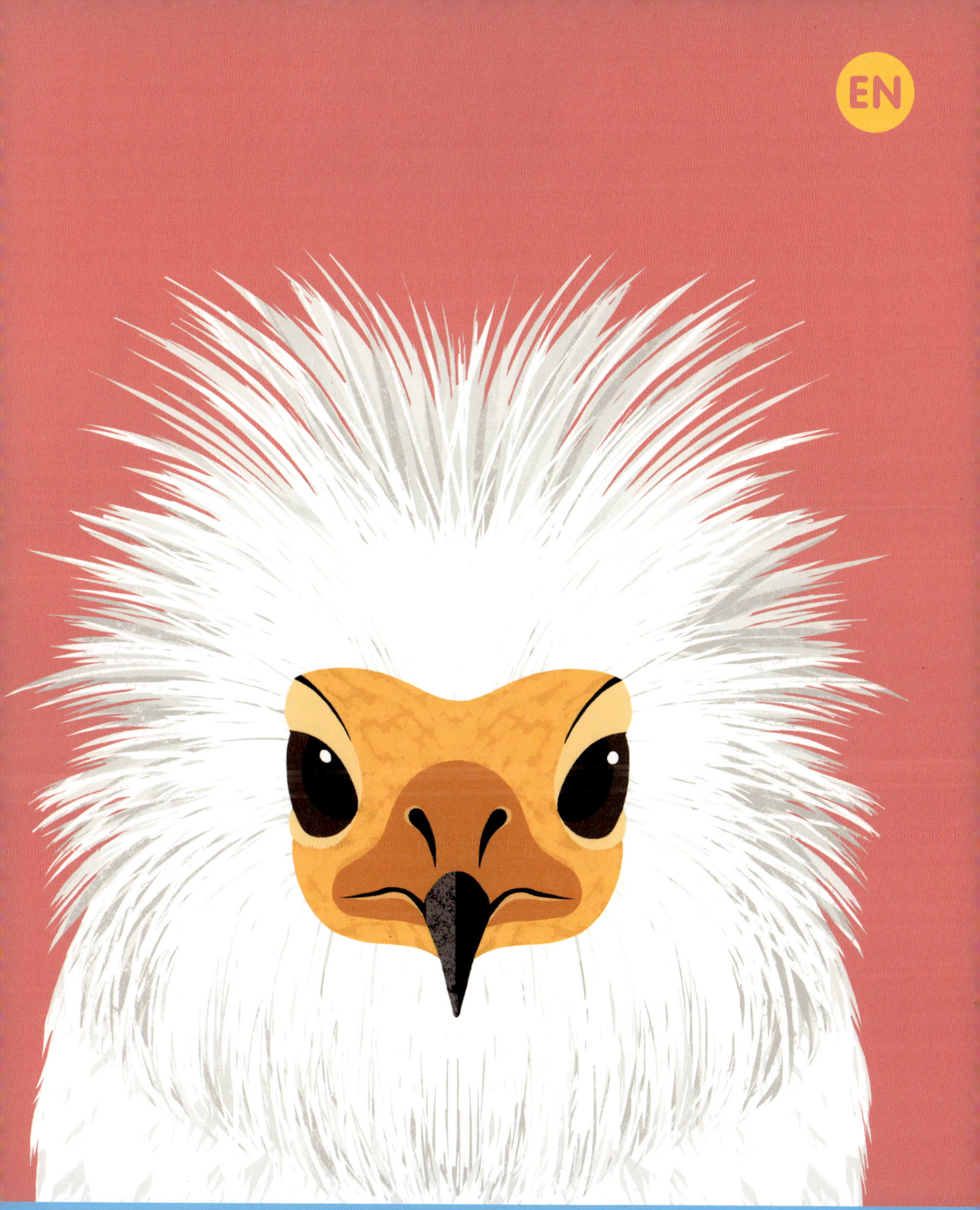

- EX Extinct
- EW Extinct in the Wild
- CR Critically Endangered
- EN Endangered
- VU Vulnerable
- NT Near Threatened
- LC Least Concern

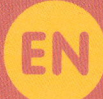

EGYPTIAN VULTURE
ENDANGERED SPECIES

European ground squirrel,
I don't need a tree house

나무집은 필요없어요, 유럽 땅다람쥐

Size 크기	20~23cm / 0.2~0.3kg	**Lifespan** 평균수명	8~10 years
Diet 식성	Omnivore 잡식	**Habitat** 서식지	Grassland 초원
Behavior 행동	Group living 무리생활	**Distribution** 분포지역	Austria, Bulgaria, Czechia, Greece, Hungary 오스트리아, 불가리아, 체코, 그리스, 헝가리
Reproduction 번식	Average 5~8 평균 5~8마리	**Conservation status** 멸종위기등급	Endangered (EN) 위기종

European ground squirrels live in short grass or meadows in order to dig burrows, and in the winter they block the entrance of their burrows with dry leaves and hibernate.
유럽땅다람쥐는 굴을 파기 위해 짧은 풀밭이나 목초지에서 살며, 겨울에는 굴 입구를 마른 잎으로 막고 겨울잠을 잡니다.

EUROPEAN GROUND SQUIRREL
ENDANGERED SPECIES

- EX Extinct
- EW Extinct in the Wild
- CR Critically Endangered
- EN Endangered
- VU Vulnerable
- NT Near Threatened
- LC Least Concern

Gerenuk,
i'm cool without water

물 없이도 괜찮아, 게레눅

Size 크기	140~160cm / 30~60kg
Diet 식성	Herbivore 초식
Behavior 행동	Group living 무리생활
Reproduction 번식	Average 1 평균 1마리
Lifespan 평균수명	10~12 years
Habitat 서식지	Desert, scrubland 사막, 관목지
Distribution 분포지역	Tanzania, Kenya, southern Somalia 탄자니아, 케냐, 소말리아 남부지역
Conservation status 멸종위기등급	Near threatened (NT) 준위협종

Gerenuk doesn't necessitate water that they can even live without water for a life time.
게레눅은 일생동안 물을 마시지 않고도 살 수 있을 만큼 물을 거의 필요로 하지 않습니다.

WE'VE BEEN SAVED—AMERICAN BISON

버팔로라는 명칭으로 알려진 아메리카 들소는 19세기 초 북미에서의 야생 개체 수가 6,000만마리로 기록된 역사상 가장 큰 동물 집단 중 하나였습니다. 아메리카 원주민에게 아메리카 들소는 고기와 가죽, 사냥 도구를 얻을 수 있는 주요 수단이자 영적인 존재로 여겨지는 동물 이었기에 필요이상의 사냥을 하지 않았으며 개체 수에 영향을 주지 않고 조화롭게 살 수 있었습니다.

20세기 초 유럽의 개척자들이 영토를 차지하기 위해 선택한 방식은 원주민들의 삶과 밀접한 관계가 있는 아메리카 들소를 없애는 것이었습니다. 대량 학살로 인해 1889년까지 북아메리카에서 살아남은 들소는 1,000마리 미만으로, 한 세기만에 수천만 마리에서 수천 마리로 수가 줄어들며 멸종 위기에 처하게 되었습니다.

아메리카 들소의 멸종을 막고자 1905년 미국 들소 협회 (American bison Society)가 설립되었습니다. 1907년 정부와 협회가 힘을 합쳐 동물원에 보호되고 있던 15마리 아메리카 들소를 야생 보호 구역에 방사하고, 서식지 생태 복원 사업을 시작했습니다. 몬태나 주에서는 대초원 보호 구역을 만들고 생태를 관리하고 있으며, 국립 야생 동물 보호 협회는 초원지대에 살고 있는 아메리카 원주민에게 아메리카 들소를 기부하거나 부족 지도자들과 협력하여 원주민과 아메리카 들소간의 전통 문화를 되살려 부족과 함께 살아갈 수 있도록 지원하고 있습니다.

이 외에도 미국 전역의 국립 공원에서 아메리카 들소를 볼 수 있는데, 특히 옐로 스톤 국립공원은 5,000여 마리의 들소가 살고 있는 미국 최대의 들소 서식지입니다. 현재는 정부와 단체들의 노력으로 개체 수가 꾸준히 증가하여 북미에서만 약 35만 마리가 서식중인 것으로 추정됩니다. 아메리카들소는 야생동물 종 보전에 성공한 대표적인 사례로 여겨집니다.

멸종위기로부터 벗어난 동물들 - 아메리카 들소

The American bison, also known as the buffalo, was one of the largest animal populations in history, with a recorded wild population of 60 million individuals in North America during the early 19th century. The Native Americans considered the bison to be a spiritual animal and a crucial source of meat, leather, and hunting tools, and thus they hunted them in harmony with nature without causing any significant impact on the population.

However, during the early 20th century, European settlers, who sought to claim the land, chose to eliminate the bison, which was closely intertwined with the Native Americans' way of life. As a result of the massive slaughter, the bison population dwindled to less than 1,000 individuals by 1889, from tens of millions just a century before, placing them in danger of extinction.

To prevent the extinction of the American bison, the American Bison Society was established in 1905. In 1907, the government and the society collaborated to reintroduce 15 bison from zoos to a protected area and initiated habitat restoration projects. In Montana, the Great Plains Conservation Area was created to manage the ecology, while the National Wildlife Federation donates bison to Native Americans living in the prairie region, supporting the revival of the bison's cultural significance among tribes and their coexistence with the bison.

In addition, American bison can be observed in national parks throughout the United States. Yellowstone National Park, in particular, is the largest bison habitat in the country, with over 5,000 individuals. Thanks to government and organization efforts, the bison population has steadily increased to an estimated 350,000 individuals in North America. The American bison is considered a successful case of wildlife conservation.

Colobus,
titanium stomach
강철 위장, 콜로부스

Size 크기	45~72cm / 5.4~20kg
Diet 식성	Omnivore 잡식
Behavior 행동	Group living 무리생활
Reproduction 번식	Average 1 평균 1마리
Lifespan 평균수명	20 years
Habitat 서식지	Forest 숲
Distribution 분포지역	Kenya 케냐
Conservation status 멸종위기등급	Vulnerable (VU) 취약종

Colobus has a strong stomach that can even digest a venomous leaf.
콜로부스는 독이 들어있는 나뭇잎을 소화할 만큼 강한 위장을 갖고있습니다.

COLOBUS
VULNERABLE SPECIES

EX Extinct
EW Extinct in the Wild
CR Critically Endangered
EN Endangered
VU Vulnerable
NT Near Threatened
LC Least Concern

Diana Monkey,
goddess of the moon
달의 여신, 다이아나 원숭이

Size 크기	40~55cm / 4~7kg
Diet 식성	Omnivore 잡식
Behavior 행동	Group living 무리생활
Reproduction 번식	Average 1 평균 1마리
Lifespan 평균수명	20 years
Habitat 서식지	Forest 숲
Distribution 분포지역	Côte d'Ivoire, Guinea, Liberia, Sierra Leone 코트디부아르, 기니, 라이베리아, 시에라리온
Conservation status 멸종위기등급	Endangered (EN) 위기종

Diana monkey was named after the Roman goddess of the moon, 'Diana', due to the white crescent-shaped fur on its forehead.
다이아나 원숭이는 이마에 초승달 모양의 흰 털 때문에 그리스 로마신화 달의 여신 '다이아나'로부터 이름을 따오게 되었습니다.

DIANA MONKEY
ENDANGERED SPECIES

EX Extinct
EW Extinct in the Wild
CR Critically Endangered
EN Endangered
VU Vulnerable
NT Near Threatened
LC Least Concern

Western red colobus,
Aliens in the Forest
숲속의 외계인, 서부 붉은 콜로버스

Size 크기	45~67cm / 6~12kg	**Habitat** 서식지	Forest, Savanna 숲, 사바나
Diet 식성	Herbivore 초식	**Distribution** 분포지역	Côte d'Ivoire, Gambia, Senegal, Guinea 코트디부아르, 감비아, 세네갈, 기니
Behavior 행동	Group living 무리생활	**Conservation status** 멸종위기등급	Endangered (EN) 위기종
Reproduction 번식	Average 1 평균 1마리		

WESTERN RED COLOBUS
ENDANGERED SPECIES

EX Extinct
EW Extinct in the Wild
CR Critically Endangered
EN Endangered
VU Vulnerable
NT Near Threatened
LC Least Concern

Gray wolf,
romantic guy
로맨틱 가이, 회색늑대

Size 크기	87~130cm / 23~80kg	Lifespan 평균수명	10 years
Diet 식성	Carnivore 육식	Habitat 서식지	Mountain, Coniferous forest 산, 침엽수림
Behavior 행동	Group living 무리생활	Distribution 분포지역	Northern Africa, Eurasia 북아프리카, 유라시아 등
Reproduction 번식	Average 6 평균 6마리	Conservation status 멸종위기 등급	Least concern (LC) 관심대상종

A male wolf mates with only one female during its lifetime.
수컷 늑대는 일생 동안 오직 한 마리의 암컷과 짝짓기를 합니다.

GRAY WOLF
LEAST CONCERN SPECIES

- EX Extinct
- EW Extinct in the Wild
- CR Critically Endangered
- EN Endangered
- VU Vulnerable
- NT Near Threatened
- LC Least Concern

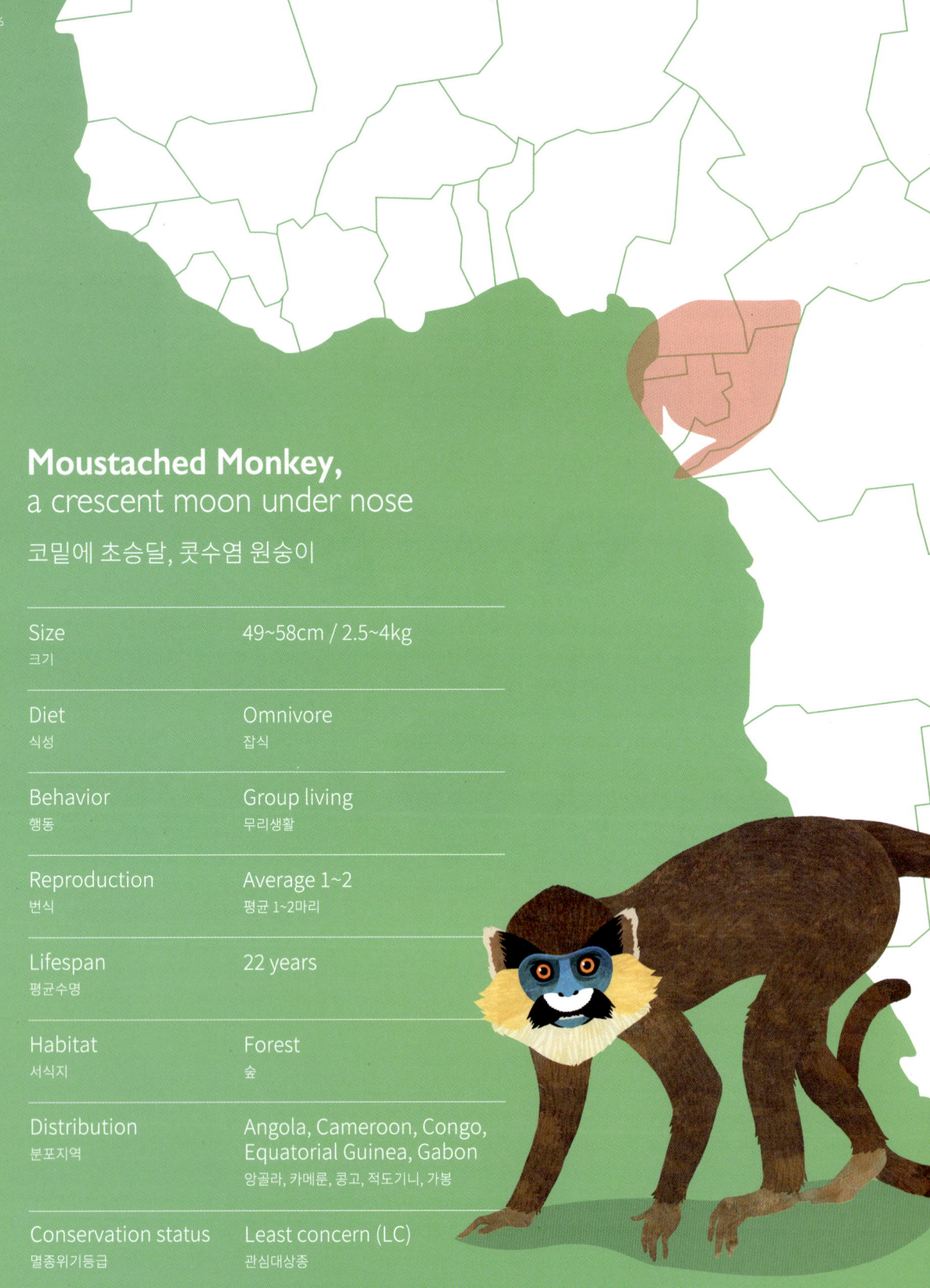

Moustached Monkey,
a crescent moon under nose
코밑에 초승달, 콧수염 원숭이

Size 크기	49~58cm / 2.5~4kg
Diet 식성	Omnivore 잡식
Behavior 행동	Group living 무리생활
Reproduction 번식	Average 1~2 평균 1~2마리
Lifespan 평균수명	22 years
Habitat 서식지	Forest 숲
Distribution 분포지역	Angola, Cameroon, Congo, Equatorial Guinea, Gabon 앙골라, 카메룬, 콩고, 적도기니, 가봉
Conservation status 멸종위기등급	Least concern (LC) 관심대상종

The mustached monkey is a primary target of illegal poaching activities, and in West Africa, they are also sold as meat.
콧수염원숭이는 불법 밀렵 활동의 주요 대상이며, 서아프리카에서는 고기로 판매되기도 합니다.

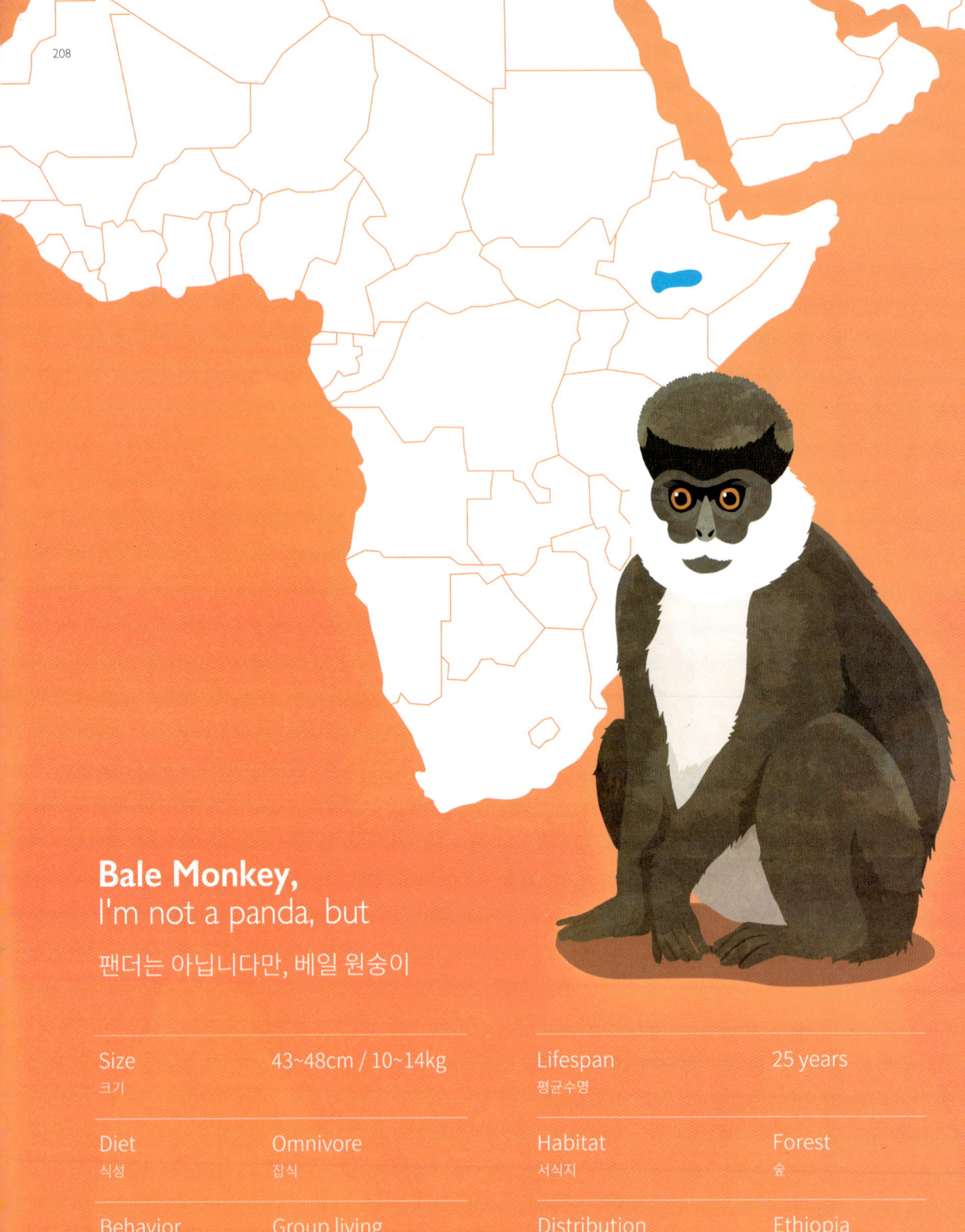

Bale Monkey,
I'm not a panda, but
팬더는 아닙니다만, 베일 원숭이

Size 크기	43~48cm / 10~14kg	**Lifespan** 평균수명	25 years
Diet 식성	Omnivore 잡식	**Habitat** 서식지	Forest 숲
Behavior 행동	Group living 무리생활	**Distribution** 분포지역	Ethiopia 에티오피아
Reproduction 번식	Average 1~2 평균 1~2마리	**Conservation status** 멸종위기등급	Vulnerable (VU) 취약종

The Bale monkey inhabits the bamboo forests of the Bale Mountains and enjoys eating bamboo.
베일 산맥의 대나무 숲에서 서식하며, 대나무 먹는 것을 좋아합니다.

VU

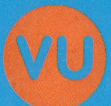

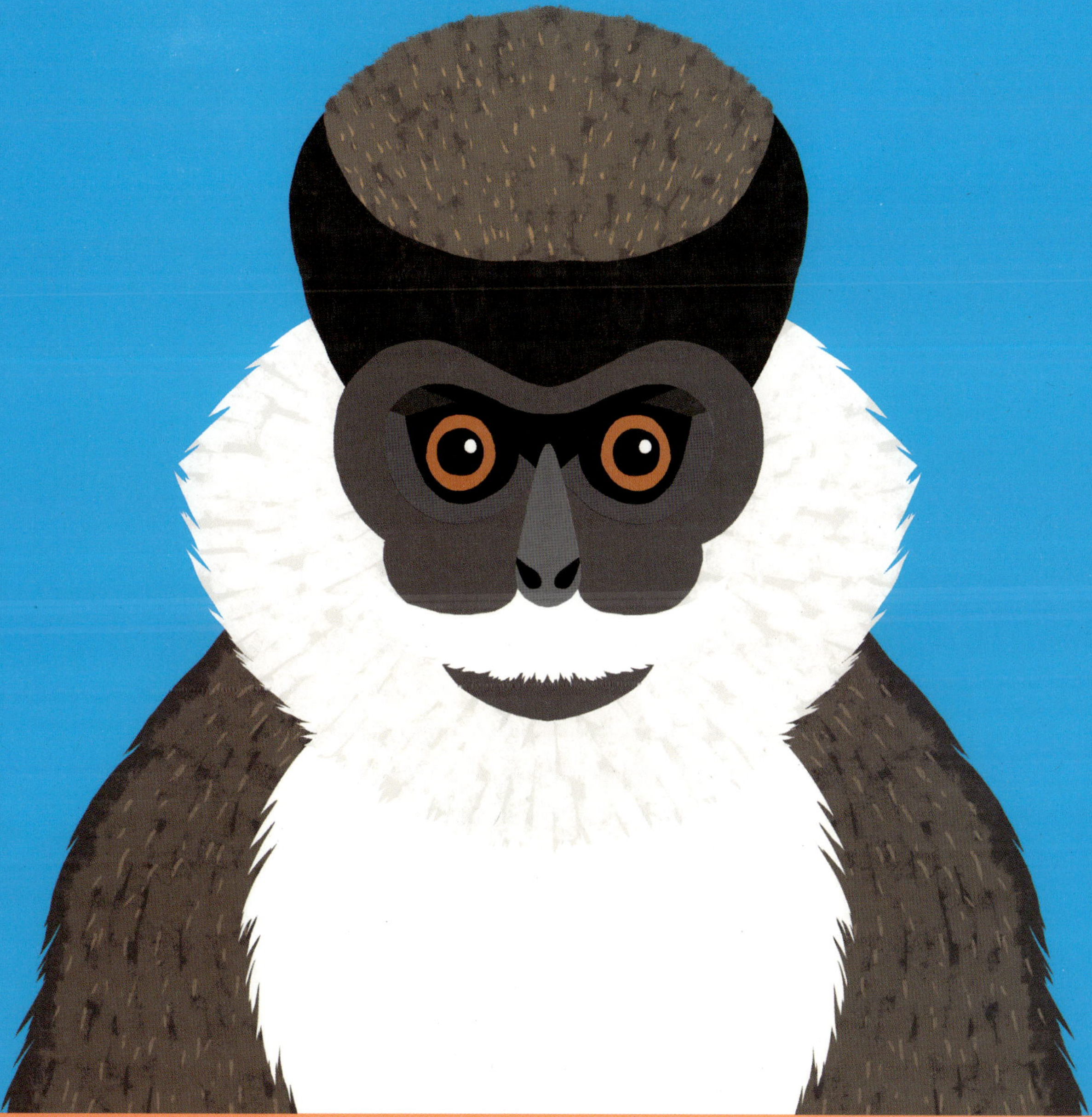

BALE MONKEY
VULNERABLE SPECIES

- EX Extinct
- EW Extinct in the Wild
- CR Critically Endangered
- EN Endangered
- VU **Vulnerable**
- NT Near Threatened
- LC Least Concern

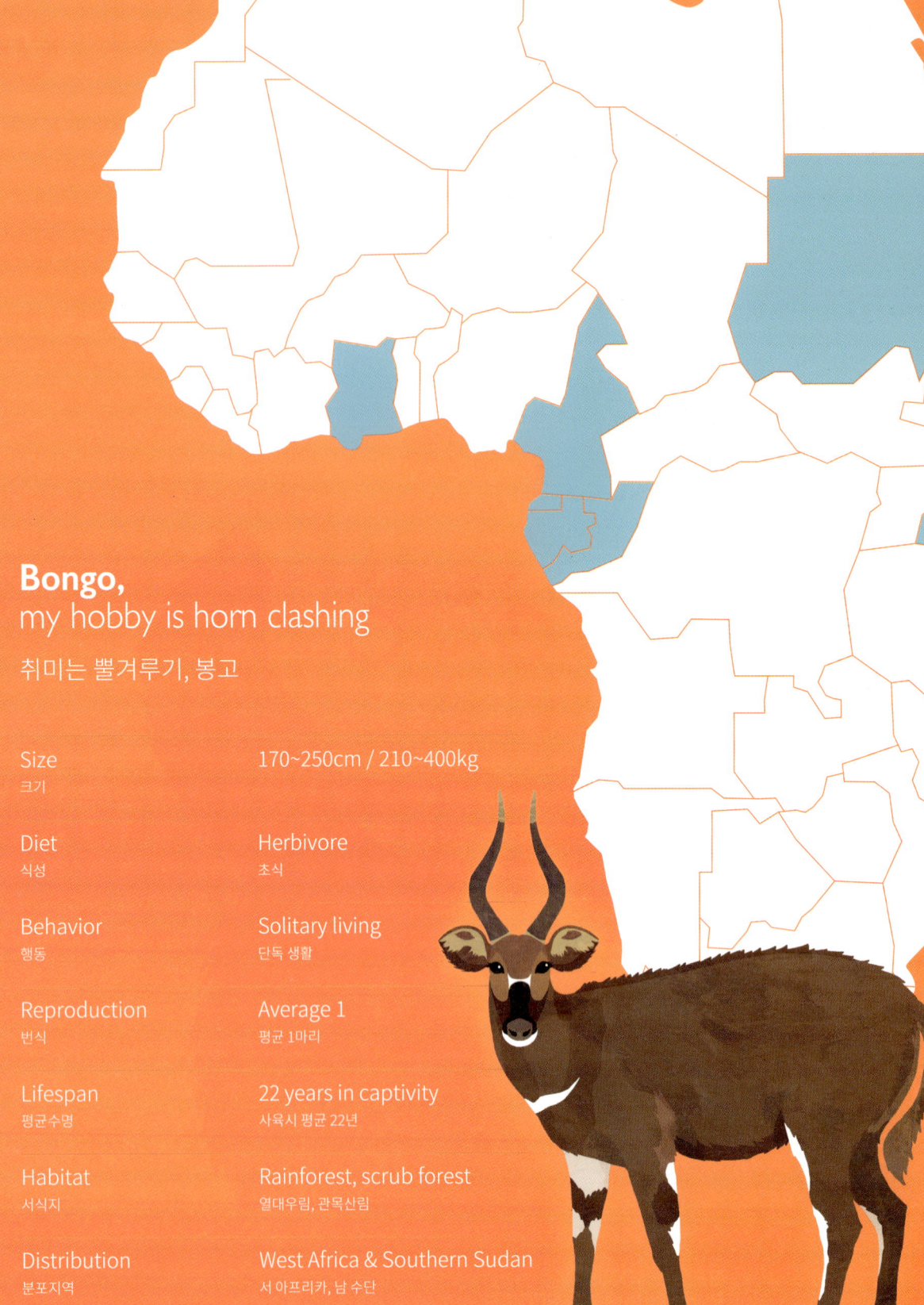

Bongo,
my hobby is horn clashing
취미는 뿔겨루기, 봉고

Size 크기	170~250cm / 210~400kg
Diet 식성	Herbivore 초식
Behavior 행동	Solitary living 단독 생활
Reproduction 번식	Average 1 평균 1마리
Lifespan 평균수명	22 years in captivity 사육시 평균 22년
Habitat 서식지	Rainforest, scrub forest 열대우림, 관목산림
Distribution 분포지역	West Africa & Southern Sudan 서 아프리카, 남 수단
Conservation status 멸종위기 등급	Near threatened (NT) 준위협종

WE'VE BEEN SAVED— ARABIAN ORYX

아라비아 오릭스가 멸종 위기종이 된 주요 원인은 과도한 사냥 및 서식지 파괴 때문이었습니다. 아라비아 반도에서 석유가 발견되면서 유동인구가 많아졌고, 1930년대 부터 아라비아의 왕자들과 석유회사 직원들 사이에서 자동차와 소총을 이용한 아라비아 오릭스 사냥이 인기를 끌기 시작하면서 그 규모가 점점 커져갔습니다. 또한 국제 사회에서 석유의 경제적 가치가 높아지면서 오만 정부는 아라비아 오릭스 서식지 보호 지역을 90% 가까이 축소하고 석유 탐사를 감행했습니다. 결국 1994년 유네스코 세계 유산에 등재됐던 아라비아 오릭스 보호 지역은 2007년 세계유산 자격을 박탈 당했습니다. 또한 과도한 방목 및 가뭄에 의한 서식지 황폐화, 사우디 아라비아에서 발생한 1999~2008년 대 가뭄으로 인해 많은 개체가 사라졌고, 서식지를 잃었습니다. 이후 아라비아 오릭스의 멸종을 막기 위해 오만, 사우디아라비아, 이스라엘 등 다양한 국가에서 적극적인 보호 정책과 번식 프로그램을 시작했습니다. 1962년 피닉스 동물원은 '오릭스 작전'이라고 불리는 번식 프로그램을 진행했습니다. 9마리 개체로 시작해 240마리까지 번식을 시키는데 성공했고, 더 많은 번식을 위해 다른 여러 동물원과 공원으로 보내는 작업을 통해 번식의 규모를 점차 키워갔습니다.

지속적인 보호와 종 보전 프로그램을 진행한 결과, 2020년 12월 기준 1,200마리 이상의 야생 개체가 있다고 추정되며, 전 세계적으로 보호중인 개체를 포함해 6,000~7,000마리가 있는 것으로 추정됩니다. 아라비아 오릭스는 국제 혈통서로 관리되며 카타르, 사우디 아라비아, 바레인, 아랍에미리트 내 일부 개체는 국가 보호 구역에서 서식합니다.

이러한 각 국가들의 노력으로 아라비아 오릭스의 개체 수는 꾸준히 늘어 났고 IUCN은 2011년 멸종위기등급을 야생 절멸종(EW)에서 취약종(VU)으로 분류하게 됩니다. IUCN이 야생에서 멸종된 후 취약종으로 등급을 낮춘 최초의 사례이기도 합니다.

멸종위기로부터 벗어난 동물들 - 아라비아 오릭스

The Arabian Oryx was driven to the brink of extinction primarily due to excessive hunting and habitat destruction. With the discovery of oil in the Arabian Peninsula, there was an increase in human population, and in the 1930s, hunting the Arabian Oryx with cars and rifles became popular among Arabian princes and oil company employees, leading to an increase in its hunting. Moreover, as the economic value of oil increased in the international community, the Omani government reduced the protected areas of the Arabian Oryx habitat to nearly 90% and conducted oil exploration. As a result, the Arabian Oryx Protection Area, which was listed as a UNESCO World Heritage site in 1994, was stripped of its world heritage status in 2007. Additionally, habitat degradation due to overgrazing and drought, and the 1999-2008 drought in Saudi Arabia caused the disappearance of many individuals and the loss of habitat.

Subsequently, to prevent the extinction of the Arabian Oryx, various countries such as Oman, Saudi Arabia, and Israel started active protection policies and breeding programs. In 1962, the Phoenix Zoo initiated a breeding program called the "Oryx Project", starting with 9 individuals and succeeding in breeding up to 240 individuals, gradually increasing the scale of breeding through sending them to other zoos and parks.

As a result of continuous protection and species conservation programs, it is estimated that there are over 1,200 wild individuals as of December 2020, and it is estimated that there are 6,000-7,000 individuals worldwide, including those under protection. The Arabian Oryx is managed as an international pedigree and some individuals in Qatar, Saudi Arabia, Bahrain, and the United Arab Emirates live in protected areas.

Thanks to the efforts of each country, the number of Arabian Oryx individuals has steadily increased, and in 2011, the IUCN downgraded its conservation status from "Extinct in the Wild" (EW) to "Vulnerable" (VU), the first case of the IUCN downgrading from EW status to VU status.

Caracal,
elf eared cat

요정의 귀를 가진 고양이, 카라칼

Size 크기	80~125cm / 8~19kg	Lifespan 평균수명	12 years
Diet 식성	Carnivore 육식	Habitat 서식지	Savanna, semidesert 사바나, 반사막
Behavior 행동	Solitary living 단독 생활	Distribution 분포지역	Central Asia, South Africa 중앙아시아, 남아프리카
Reproduction 번식	Average 3 평균 3마리	Conservation status 멸종위기 등급	Least concern (LC) 관심대상종

Manatee,
a motherly mermaid
모성애 넘치는 인어, 매너티

Size 크기	2.5~4.6m / 0.3~1.6t	Lifespan 평균수명	60 years
Diet 식성	Seaweed 해조류	Habitat 서식지	Marine Neritic 얕은 바다
Behavior 행동	Group living 무리생활	Distribution 분포지역	Western Africa 서부 아프리카
Reproduction 번식	Average 1 평균 1마리	Conservation status 멸종위기등급	Vulnerable (VU) 취약종

Manatees, also known as mermaids due to their human-like behavior of carrying their young and nursing them, breastfeed their offspring for nearly two years.
어린 새끼를 안고 젖을 먹이는 모습이 사람을 닮아 인어라 불리며 2년 가까이 모유 수유를 합니다.

MANATEE
VULNERABLE SPECIES

EX Extinct
EW Extinct in the Wild
CR Critically Endangered
EN Endangered
VU Vulnerable
NT Near Threatened
LC Least Concern

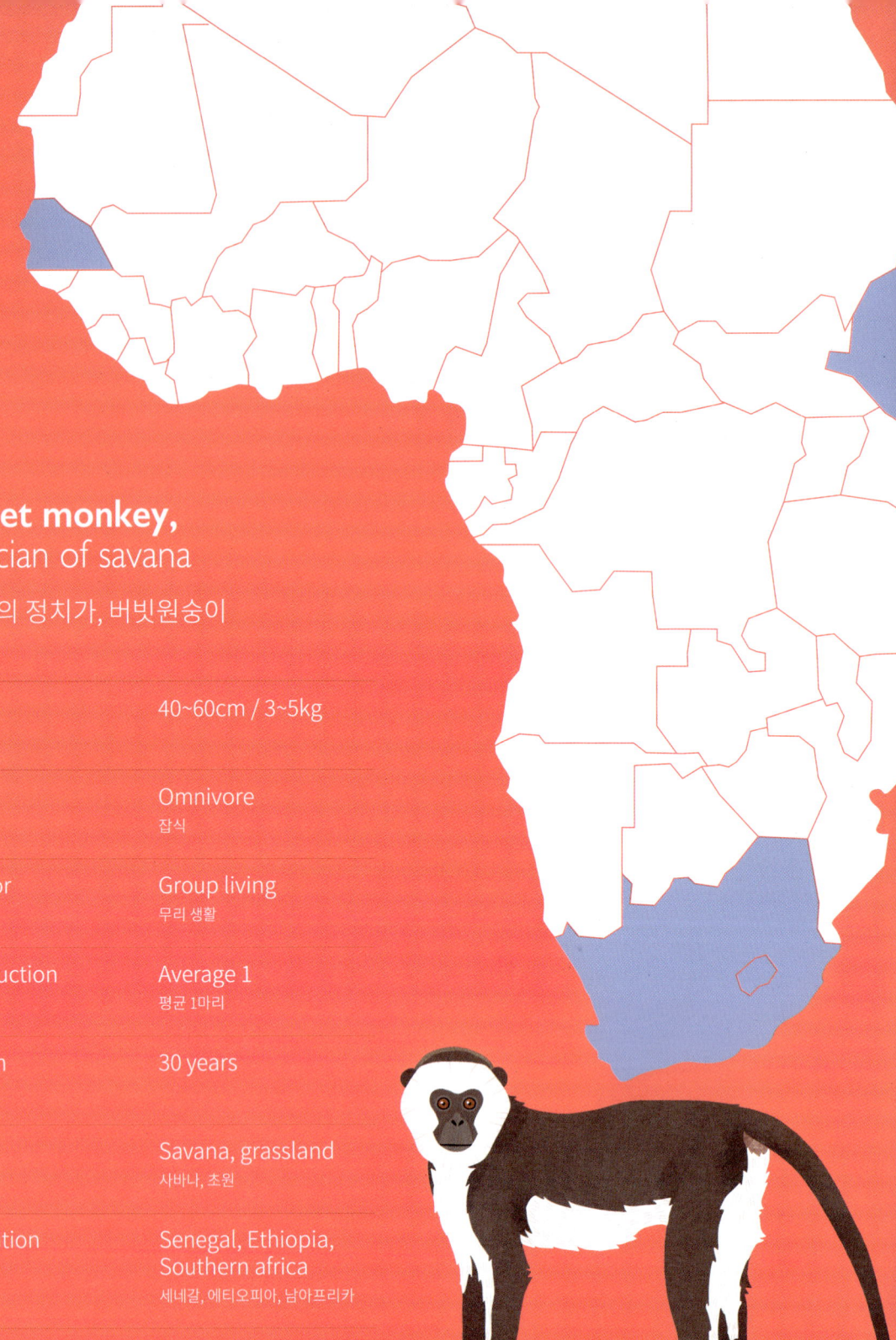

Vervet monkey,
politician of savana
사바나의 정치가, 버빗원숭이

Size 크기	40~60cm / 3~5kg
Diet 식성	Omnivore 잡식
Behavior 행동	Group living 무리 생활
Reproduction 번식	Average 1 평균 1마리
Lifespan 평균수명	30 years
Habitat 서식지	Savana, grassland 사바나, 초원
Distribution 분포지역	Senegal, Ethiopia, Southern africa 세네갈, 에티오피아, 남아프리카
Conservation status 멸종위기 등급	Least concern (LC) 관심대상종

In a society of Vervet monkeys, there is a strict social hierarchy.
Even grownup monkeys should obey the younger ones that have a higher social rank.
성인 원숭이도 높은 사회적 능력이 있는 청소년 원숭이 앞에선 복종해야 하는 엄격한 사회계층이 존재합니다.

Striped hyena, the lonley nomad
고독한 방랑자, 줄무늬 하이에나

Size 크기	65~80cm / 25~45kg	**Lifespan** 평균수명	20~24 years
Diet 식성	Carnivore 육식	**Habitat** 서식지	Arid mountainous region 건조한 산악 지대
Behavior 행동	Solitary living 단독생활	**Distribution** 분포지역	Northern & Eaternafrica, India southern Asia 북부 및 동부 아프리카, 인도, 남부아시아
Reproduction 번식	Average 2~3 평균 2~3마리	**Conservation status** 멸종위기등급	Near threatened (NT) 준위협종

Striped hyena is a solitary animal and wanders around irregularly in search of its prey.
줄무늬 하이에나는 단독생활을 하고 불규칙하게 떠돌며 먹이를 찾습니다.

Eurasian eagle owl, emperor of night
밤의 제왕, 수리부엉이

Size 크기	58~71cm / 2~5kg	**Lifespan** 평균수명	15 years
Diet 식성	Carnivore 육식	**Habitat** 서식지	Conigerous forest, desert 침엽수림, 사막
Behavior 행동	Solitary living 단독 생활	**Distribution** 분포지역	North Africa, Europe, Asia 북아프리카, 유럽, 아시아
Reproduction 번식	Average 3 평균 3마리	**Conservation status** 멸종위기 등급	Least concern (LC) 관심대상종

EURASIAN EAGLE OWL
LEAST CONCERN SPECIES

EX Extinct
EW Extinct in the Wild
CR Critically Endangered
EN Endangered
VU Vulnerable
NT Near Threatened
LC Least Concern

Atlantic Puffin,
animation star

애니메이션 단골 출연, 코뿔바다오리

Size 크기	43~48cm / 10~14kg	**Lifespan** 평균수명	25 years
Diet 식성	Omnivore 잡식	**Habitat** 서식지	Forest 숲
Behavior 행동	Group living 무리생활	**Distribution** 분포지역	Ethiopia 에티오피아
Reproduction 번식	Average 1~2 평균 1~2마리	**Conservation status** 멸종위기등급	Vulnerable (VU) 취약종

Atlantic puffin has appeared in various animations such as Puffin Rock, Happy Feet, and more.
행복한 퍼핀가족, 해피 피트 등 여러 애니메이션에 등장하였습니다.

ATLANTIC PUFFIN
VULNERABLE SPECIES

EX Extinct
EW Extinct in the Wild
CR Critically Endangered
EN Endangered
VU Vulnerable
NT Near Threatened
LC Least Concern

WE'VE BEEN SAVED— BALD EAGLE

미국의 국조로 알려진 흰머리 독수리는 1950년대 미국 48개 주에 단 412쌍만이 남았다는 조사와 함께 1967년 멸종위기종이 되었습니다. 1963년 부터 1984년까지 National Wildlife Health Center 에서 진행된 흰머리 독수리의 사망 원인 분석 결과 1,428마리의 독수리 중 23%는 교통 사고, 22%는 총상, 11%는 독극물 중독, 9%는 감전, 5%는 포획인 것으로 확인되어 사망 원인의 대부분이 인간에 의한 것으로 밝혀졌습니다.

1940년 미국 의회에서 승인된 '흰머리 독수리 보호법'에 의해 이 종의 상업적 포획을 포함한 모든 사냥을 금지하고, 1970년대부터 신경세포를 파괴시키며 해충을 죽이는 유기 염소계 살충제인 DDT의 사용도 금지되었습니다. 또한 미 전역에서 번식 프로그램이 진행되었는데, 대표적으로 American Eagle Foundation은 1992년부터 흰머리 독수리를 구조하고 사육하여 번식 가능한 한 쌍을 방생하는 일을 하고 있습니다. 또한 일반인들이 보호 활동에 참여할 수 있도록 행동 지침서가 발간되기도 하는 등 주요 사망 원인이었던 사항들을 점차 줄여나갔습니다. 꾸준한 보호 활동을 통해 개체 수가 증가하기 시작하여 1990년대에는 11만 마리가 서식 중인 것으로 조사되었습니다. 멸종위기 보호종으로 지정된 지 40년 만인 2007년, 미국 내 멸종위기 동물 리스트에서 제외되었습니다.

멸종위기로부터 벗어난 동물들 - 흰머리 독수리

The bald eagle, also known as the national bird of the United States, was classified as an endangered species in 1967, along with a report that there were only 412 pairs left in the lower 48 states in the 1950s. From 1963 to 1984, the National Wildlife Health Center conducted an analysis of the causes of death of bald eagles, revealing that among the 1,428 eagles studied, 23% died from collisions with vehicles, 22% from gunshot wounds, 11% from poisoning, 9% from electrocution, and 5% from trapping. It was found that the vast majority of these deaths were caused by human activities.

In 1940, the Bald Eagle Protection Act was passed by the US Congress, which prohibited all hunting of this species, including commercial trapping, and the use of the organochlorine pesticide DDT, which kills pests by destroying nerve cells, was also banned from the 1970s. Breeding programs were implemented throughout the United States, and the American Eagle Foundation has been rescuing and breeding bald eagles since 1992 to release pairs back into the wild. Guidelines for protecting the eagles were also published to encourage public participation in the conservation effort. Through these conservation efforts, the number of bald eagles began to increase, and by the 1990s, there were an estimated 110,000 individuals in the wild. After 40 years of protection as an endangered species, the bald eagle was removed from the US endangered species list in 2007.

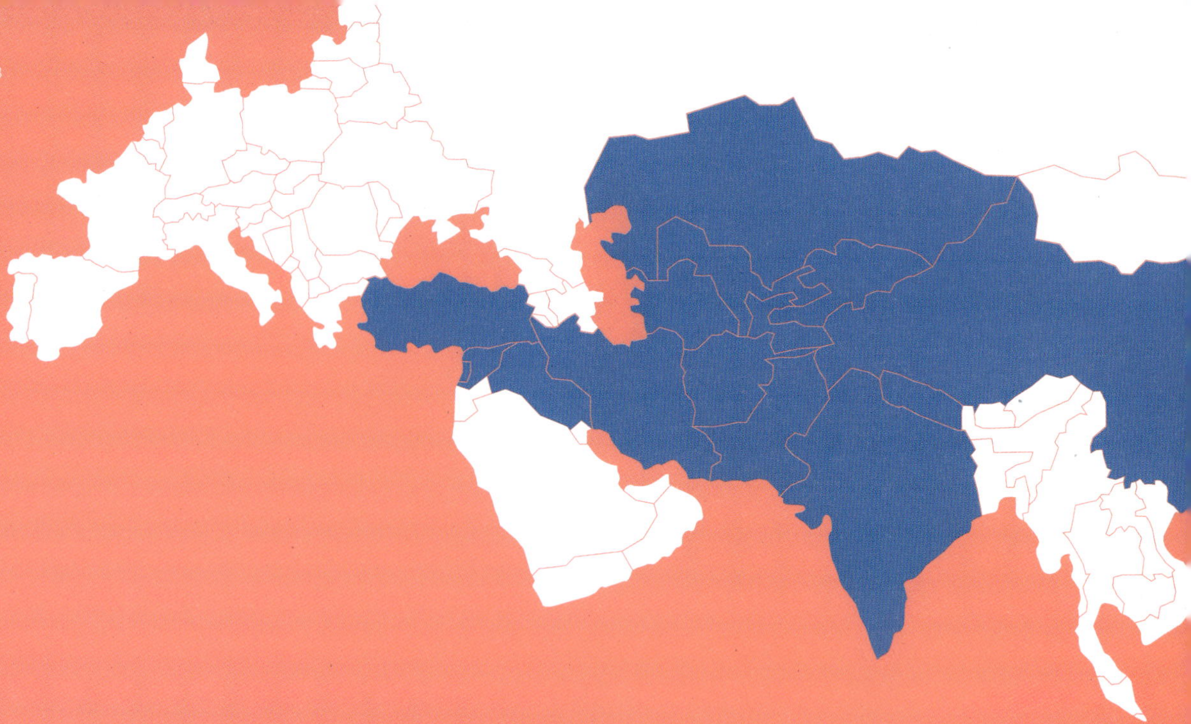

Bearded vulture,
holy eagle

신성한 독수리, 수염수리

Size 크기	94~125cm / 5~7kg	**Lifespan** 평균수명	40 years in captivity 사육시 평균 40년
Diet 식성	Carnivore 육식	**Habitat** 서식지	Mountainous region 산악지역
Behavior 행동	Solitary living 단독생활	**Distribution** 분포지역	Southern Europe, Asia 유럽 남쪽 지역, 아시아
Reproduction 번식	Average 1~3 eggs 평균 1~3개의 알	**Conservation status** 멸종위기등급	Near threatened (NT) 준위협종

In some parts of bearded vulture's habitats, they are considered as a sacred animal.
수염수리가 서식하는 일부 지역은 이 새를 신성한 존재로 여기기도 합니다.

BEARDED VULTURE
NEAR THREATENED SPECIES

EX Extinct
EW Extinct in the Wild
CR Critically Endangered
EN Endangered
VU Vulnerable
NT Near Threatened
LC Least Concern

Marbled Polecat,
the most splendid weasel

가장 화려한 족제비, 얼룩 족제비

Size 크기	29~35cm / 0.3~0.7kg	Lifespan 평균수명	8~9 years
Diet 식성	Carnivore 육식	Habitat 서식지	Shrubland, Grassland, Rocky areas 관목지대, 초원, 바위 지역
Behavior 행동	Solitary living 단독생활	Distribution 분포지역	Southeast Europe, Central Asia, China, Mongolia, Russia 유럽 남동부, 중앙아시아, 중국, 몽골, 러시아
Reproduction 번식	Average 4 평균 4마리	Conservation status 멸종위기등급	Vulnerable (VU) 취약종

The face of the spotted peacock is marked with black-and-white spots, and its back is adorned with cheetah-like spots.
얼굴엔 흑백의 반점이, 등에는 치타와 같은 반점이 있습니다.

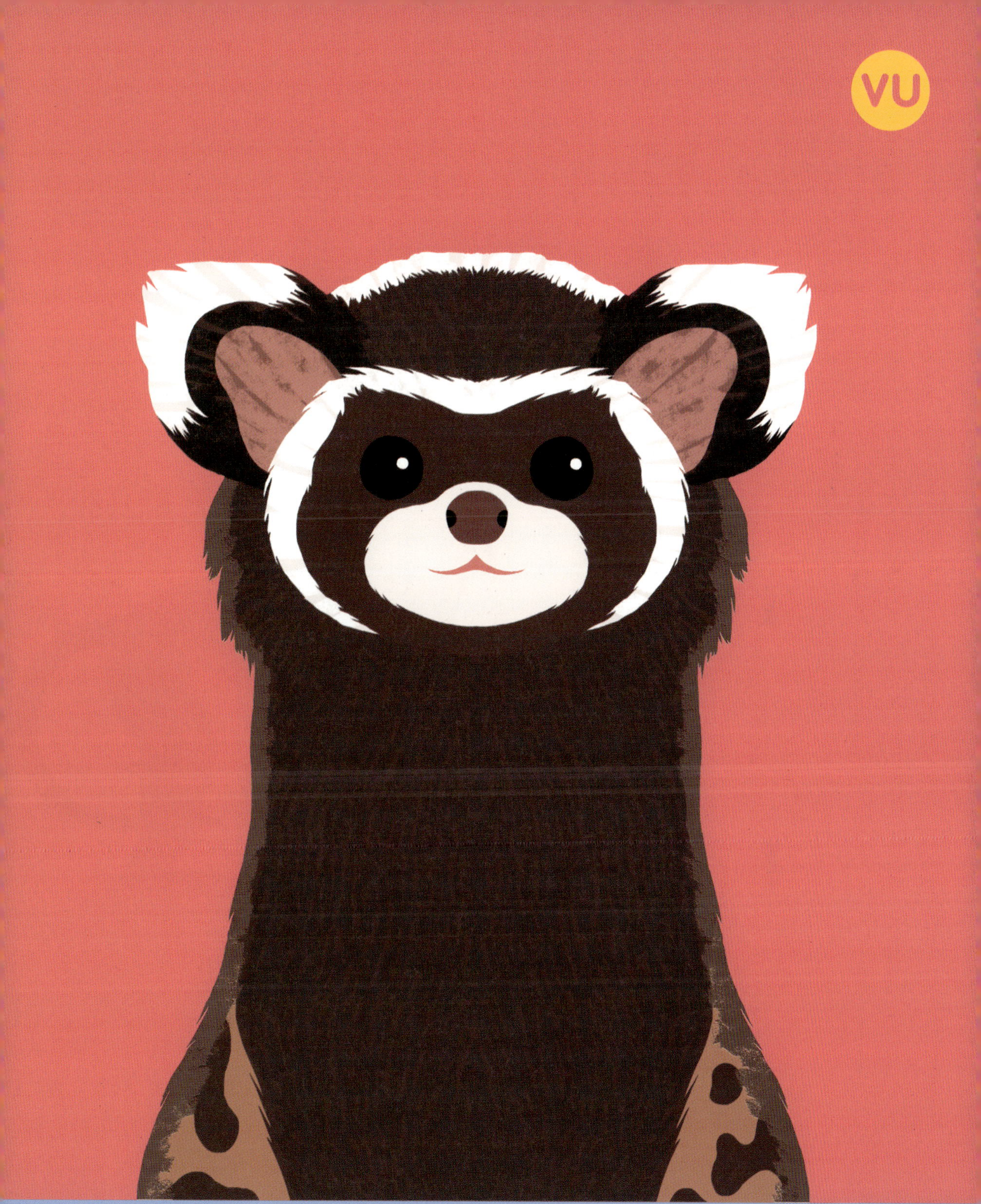

EX Extinct
EW Extinct in the Wild
CR Critically Endangered
EN Endangered
VU Vulnerable
NT Near Threatened
LC Least Concern

MARBLED POLECAT
VULNERABLE SPECIES

Malbrouck Monkey,
good communicator

소통의 달인, 말브룩 원숭이

Size 크기	30~60cm / 3.4~8kg	**Lifespan** 평균수명	15~30 years
Diet 식성	Omnivore 잡식	**Habitat** 서식지	Forest, Savanna 숲, 사바나
Behavior 행동	Group living 무리생활	**Distribution** 분포지역	Angola, Congo, Namibia, Zambia 앙골라, 콩고, 나미비아, 잠비아
Reproduction 번식	Average 1 평균 1마리	**Conservation status** 멸종위기등급	Least concern (LC) 관심대상종

Malbrouck monkey uses a wide range of sounds and gestures to communicate with other members of a group.
그룹 내 다른 구성원들과 의사소통하기 위해 다양한 소리와 제스처를 사용합니다.

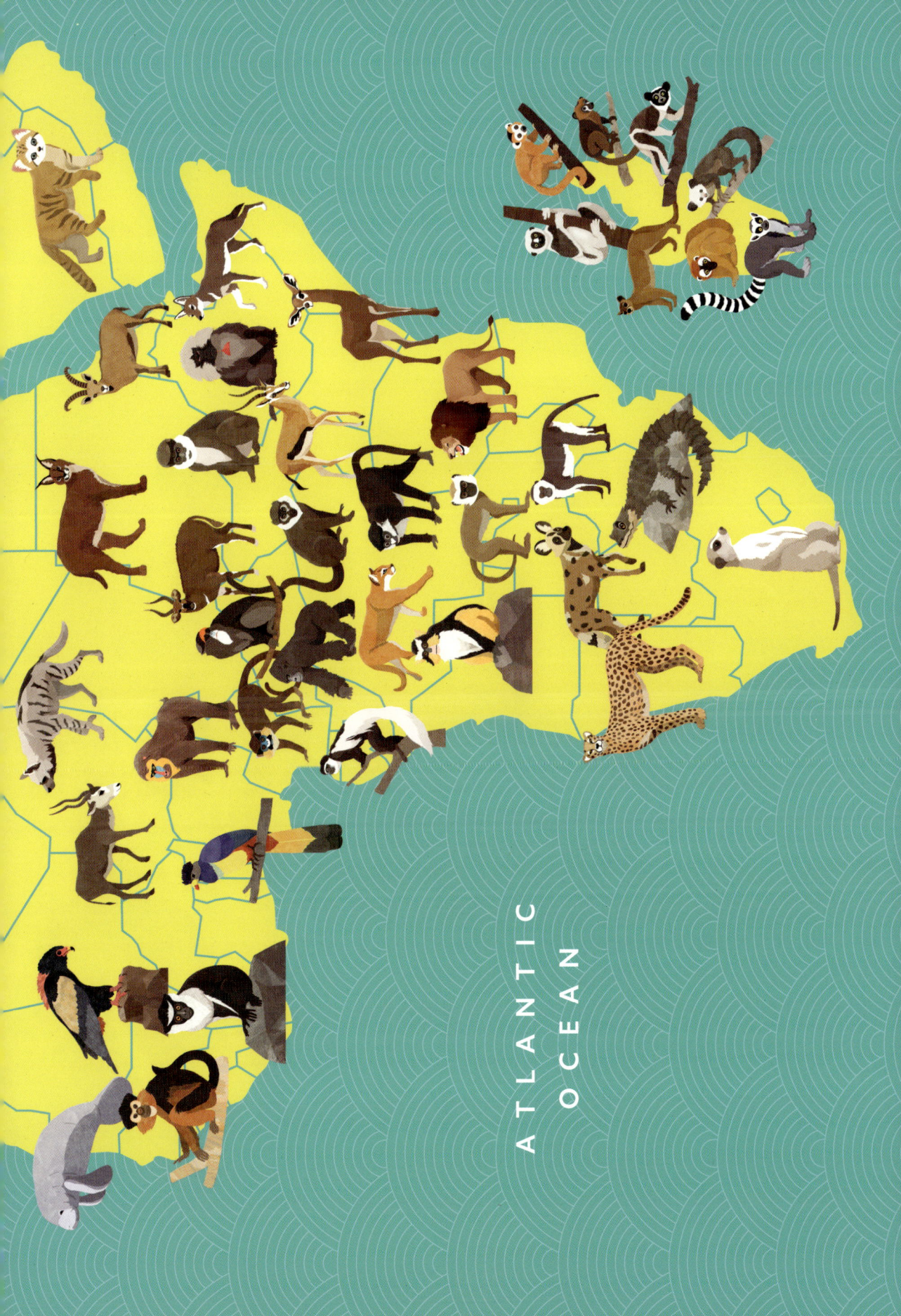
ATLANTIC OCEAN

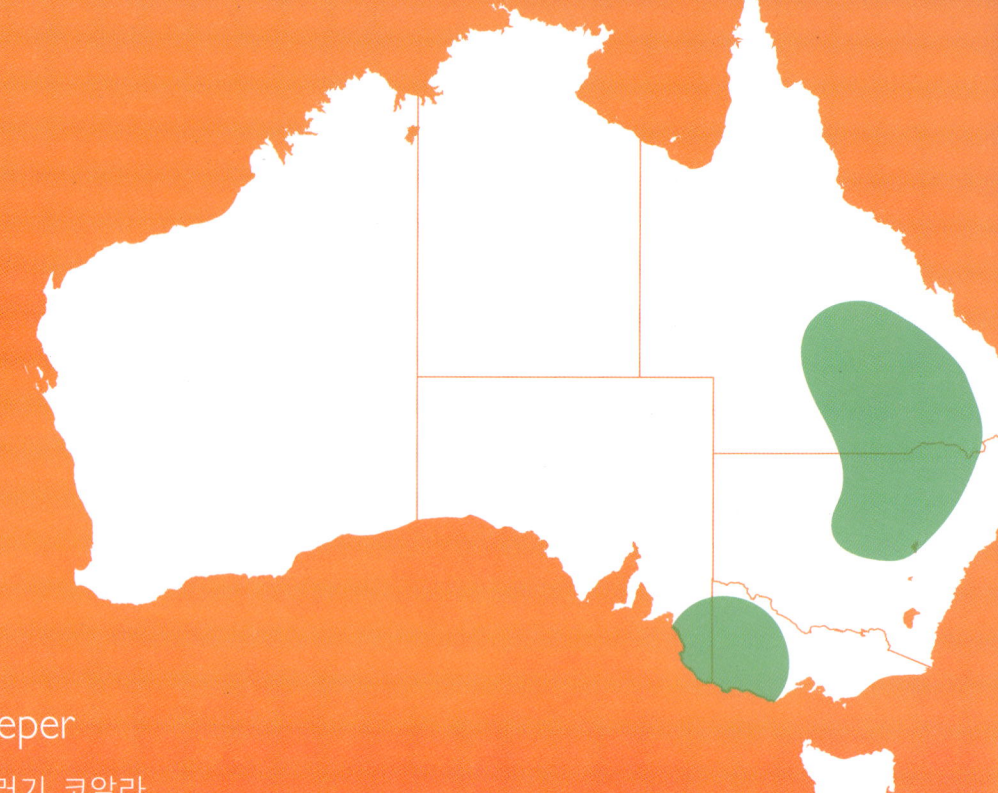

Koala,
heavy sleeper

게으른 잠꾸러기, 코알라

Size 크기	72~78cm / 5~12kg
Diet 식성	Herbivore 초식
Behavior 행동	Solitary living 단독 생활
Reproduction 번식	Average 1 평균 1마리
Lifespan 평균수명	15~20 years
Habitat 서식지	Eucalypt woodland 유칼리나무 숲
Distribution 분포지역	Eastern and Southeastern Australia 호주 동부 및 남동부
Conservation status 멸종위기 등급	Vulnerable (VU) 취약종

A koala usually sleeps 20 hours a day, it eats food the rest of the time.
하루에 보통 20시간을 자고 나머지 시간에 식사를 합니다.

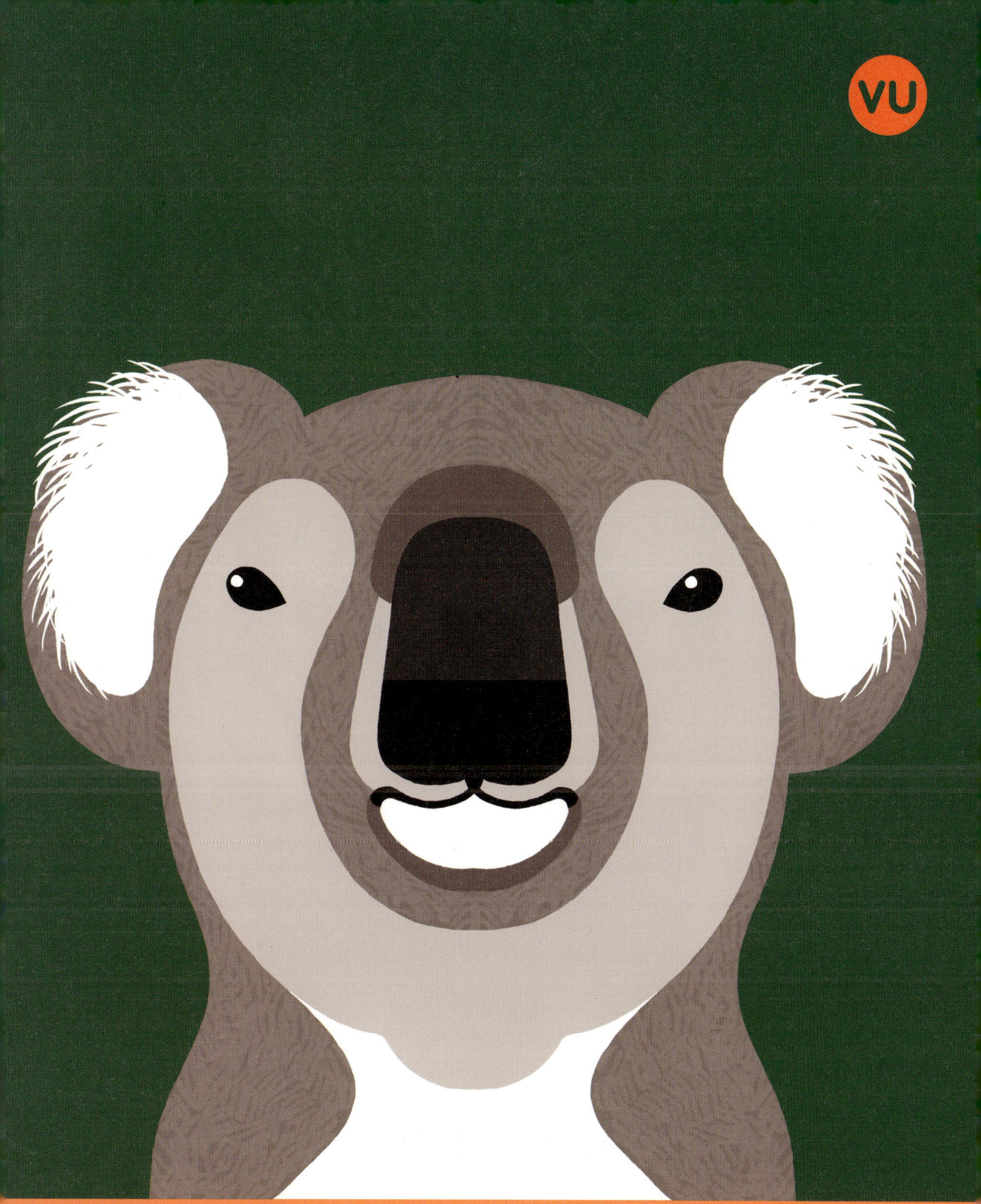

EX Extinct
EW Extinct in the Wild
CR Critically Endangered
EN Endangered
VU **Vulnerable**
NT Near Threatened
LC Least Concern

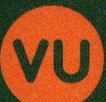

KOALA
VULNERABLE SPECIES

Siberian Flying Squirrel,
a shy aviator

수줍은 비행사, 하늘 다람쥐

Size 크기	13~20cm / 80~150g	**Lifespan** 평균수명	5 years
Diet 식성	Omnivore 잡식	**Habitat** 서식지	Forest 숲
Behavior 행동	Solitary living 단독생활	**Distribution** 분포지역	Russian, China, Molgolia, Japan, Finland, etc. 러시아, 중국, 몽골, 일본, 핀란드 등
Reproduction 번식	Average 2~3 평균 2~3마리	**Conservation status** 멸종위기등급	Least concern (LC) 관심대상종

The Siberian flying squirrel is a shy and nocturnal animal, making it not easily noticed by humans.
수줍음이 많고 야행성 동물이라 사람에게 잘 발견되지 않습니다.

SIBERIAN FLYING SQUIRREL
LEAST CONCERN SPECIES

EX Extinct
EW Extinct in the Wild
CR Critically Endangered
EN Endangered
VU Vulnerable
NT Near Threatened
LC Least Concern

Platypus,
not-a-duck, nor-a-racoon

오리도 너구리도 아닌, 오리너구리

Size 크기	43~50cm / 0.7~2.4kg	**Lifespan** 평균수명	10~17 years
Diet 식성	Carnivore 육식	**Habitat** 서식지	River, stream 강, 개울
Behavior 행동	Solitary living 단독생활	**Distribution** 분포지역	Eastern Australia 호주 동부
Reproduction 번식	Average 2 평균 2마리	**Conservation status** 멸종위기등급	Near threatened (NT) 준위협종

Platypus was doubted as an imaginary animal when it was first introduced in 1882, because of its unique looks.
오리너구리는 독특한 생김새 때문에 1882년 처음 소개되었을 때에는 가상으로 만들어낸 동물이 아니냐는 의혹을 받기도 했습니다.

PLATYPUS
NEAR THREATENED SPECIES

EX Extinct
EW Extinct in the Wild
CR Critically Endangered
EN Endangered
VU Vulnerable
NT Near Threatened
LC Least Concern

Bridled nail-tail wallaby,
I'm not kangaroo

나는 캥거루가 아니에요, 고삐 발톱 꼬리 왈라비

Size 크기	100cm / 4~8kg
Diet 식성	Herbivore 초식
Behavior 행동	Group living 무리생활
Reproduction 번식	Average 3 평균 3마리
Life span in wild 평균수명	7 years
Habitat 서식지	Dense acacia shrubland 조밀한 아카시아 관목
Distribution 분포지역	Northeast Australia 호주 북동부
Conservation status 멸종위기 등급	Vulnerable (VU) 취약종

BRIDLED NAIL-TAIL WALLABY
VULNERABLE SPECIES

- EX Extinct
- EW Extinct in the Wild
- CR Critically Endangered
- EN Endangered
- VU Vulnerable
- NT Near Threatened
- LC Least Concern

Common ringtail possum,
a cat-like mouse

고양이를 닮은 쥐, 알락꼬리 주머니쥐

Size 크기	30~35cm / 0.5~1.1kg
Diet 식성	Herbivore 초식
Behavior 행동	Group living 무리생활
Reproduction 번식	Average 2 평균 2마리
Life span in wild 평균수명	5~9 years
Habitat 서식지	Forest, Savanna 숲, 사바나
Distribution 분포지역	East Australia 호주 동부
Conservation status 멸종위기 등급	Least concern (LC) 관심대상종

Greater Glidert,
coats may vary, but we're friends.
털색은 달라도 같은 친구예요, 주머니 날다람쥐

Size 크기	39~43cm / 1.3kg
Diet 식성	Herbivore 초식
Behavior 행동	Solitary living 단독생활
Reproduction 번식	Average 1 평균 1마리
Life span in wild 평균수명	15 years
Habitat 서식지	Eucalyptus Forest 유칼립투스 숲
Distribution 분포지역	South-east Australia 호주 동남부
Conservation status 멸종위기 등급	Vulnerable (VU) 취약종

Great Gliders vary in fur color and size depending on their habitat.
서식지에 따라 털 색깔과 크기가 다릅니다.

GREATER GLIDER
VULNERABLE SPECIES

- EX Extinct
- EW Extinct in the Wild
- CR Critically Endangered
- EN Endangered
- VU Vulnerable
- NT Near Threatened
- LC Least Concern

WE'VE BEEN SAVED — BROWN PELICAN

아메리카 대륙의 대서양, 태평양 연안에 서식하며 미국 루이지애나 주의 공식 마스코트로 알려져 있습니다. 1940년대 신경세포를 파괴시켜 해충을 제거하는 유기 염소계 살충제인 DDT가 광범위하게 사용됐고, 이로 인해 개체 수가 급격히 줄어들기 시작하여 멸종 위기를 맞게 됐습니다.

1972년 미국 환경 보호국(USEPA)는 1970년부터 2009년까지 미국 멸종 위기종 관리법에 따라 DDT와 같은 살충제 사용을 금지하였고, 대통령 루스벨트는 1903년 사냥꾼들로부터 브라운 펠리컨을 보호하기 위해 플로리다 펠리컨 섬에 최초의 국립 야생 동물 보호 구역을 지정했습니다. 이러한 노력에도 불구하고 브라운 펠리컨은 1963년 루이지애나 주에서 멸종 되었습니다.

1968년 루이지애나 주는 브라운 펠리컨의 종복원을 위해 플로리다 생물학자들과 협력해 1,000여 마리를 동물원으로 옮겨 관리했고, 건강한 개체들을 자연으로 방생하였습니다. 정부의 다양한 규제와 노력들로 플로리다 주를 포함한 미국 동부 및 대서양 연안을 따라 북쪽으로 개체 수가 점차 회복되었습니다.

조사에 따르면 40년간 브라운 펠리컨의 개체 수는 기존 대비 68% 증가했으며, 2009년 일부 지역에서는 멸종 위기종에서 제외되기도 했습니다. 하지만 환경 변화로 인한 먹이 감소 및 서식지 파괴로 인한 멸종 위협의 요소가 계속 생겨나고 있어 지속적인 관리를 진행 중에 있습니다.

멸종위기로부터 벗어난 동물들 - 브라운 펠리컨

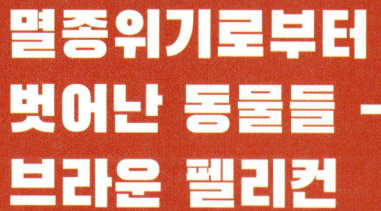

The Brown Pelican is known as the official mascot of the US state of Louisiana and inhabits the Atlantic and Pacific coasts of the American continent. In the 1940s, the organic pesticide DDT, which destroys nerve cells and eliminates pests, was widely used, causing a drastic decline in the population of Brown Pelicans and pushing them to the brink of extinction.

In 1972, the US Environmental Protection Agency (USEPA) banned the use of pesticides such as DDT under the Endangered Species Act from 1970 to 2009, and President Roosevelt designated the first national wildlife refuge on Florida's Pelican Island in 1903 to protect the Brown Pelican from hunters. Despite these efforts, the Brown Pelican became extinct in Louisiana in 1963.

In 1968, Louisiana collaborated with Florida biologists to relocate over 1,000 Brown Pelicans to zoos for conservation, and healthy individuals were released back into the wild. With various government regulations and efforts, the population gradually recovered along the US east coast and Atlantic coast, including Florida. According to surveys, the Brown Pelican population increased by 68% over 40 years compared to the previous period, and in some areas in 2009, it was even excluded from the endangered species list. However, factors such as food scarcity and habitat destruction due to environmental changes continue to threaten their survival, and ongoing conservation efforts are being implemented.

Bilby,
rabbit-like, yet not quite the same
토끼인 듯 아닌 듯, 빌비

Size 크기	30~55cm / 0.8~2.5kg		**Lifespan** 평균수명	10 years in captivity 사육시 평균 10년
Diet 식성	Omnivore 잡식		**Habitat** 서식지	Savanna, Grassland 사바나, 초원
Behavior 행동	Solitary living 단독생활		**Distribution** 분포지역	North west Australia 호주 북서부
Reproduction 번식	Average 1~3 평균 1~3마리		**Conservation status** 멸종위기등급	Vulnerable (VU) 취약종

BILBY
VULNERABLE SPECIES

EX Extinct
EW Extinct in the Wild
CR Critically Endangered
EN Endangered
VU Vulnerable
NT Near Threatened
LC Least Concern

Quokka,
grinning charmer
스마일 가이, 쿼카

Size 크기	40~54cm / 2~4kg
Diet 식성	Herbivore 초식
Behavior 행동	Group living 무리생활
Reproduction 번식	Average 1 평균 1마리
Lifespan 평균수명	5~10 years
Habitat 서식지	Scrubland, swamp 관목지, 습지
Distribution 분포지역	South western Australia 호주 남서부
Conservation status 멸종위기등급	Vulnerable (VU) 취약종

QUOKKA
VULNERABLE SPECIES

EX Extinct
EW Extinct in the Wild
CR Critically Endangered
EN Endangered
VU Vulnerable
NT Near Threatened
LC Least Concern

Black crested gibbon,
the siren gibbon

사이렌 원숭이, 검은손기번

Size 크기	43~54cm / 7~10kg	**Lifespan** 평균수명	27.5 years
Diet 식성	Herbivore 초식	**Habitat** 서식지	Deciduous forest 낙엽수림
Behavior 행동	Group living 무리생활	**Distribution** 분포지역	Hainan, Laose, Vietnam 하이난, 라오스, 베트남
Reproduction 번식	Average 1 평균 1마리	**Conservation status** 멸종위기 등급	Critically endangered (CR) 위급종

All gibbons mark and defend their area by making noises.
모든 기번은 소리로 영역을 표시하고, 방어합니다.

BLACK CRESTED GIBBON
CRITICALLY ENDANGERED SPECIES

EX Extinct
EW Extinct in the Wild
CR Critically Endangered
EN Endangered
VU Vulnerable
NT Near Threatened
LC Least Concern

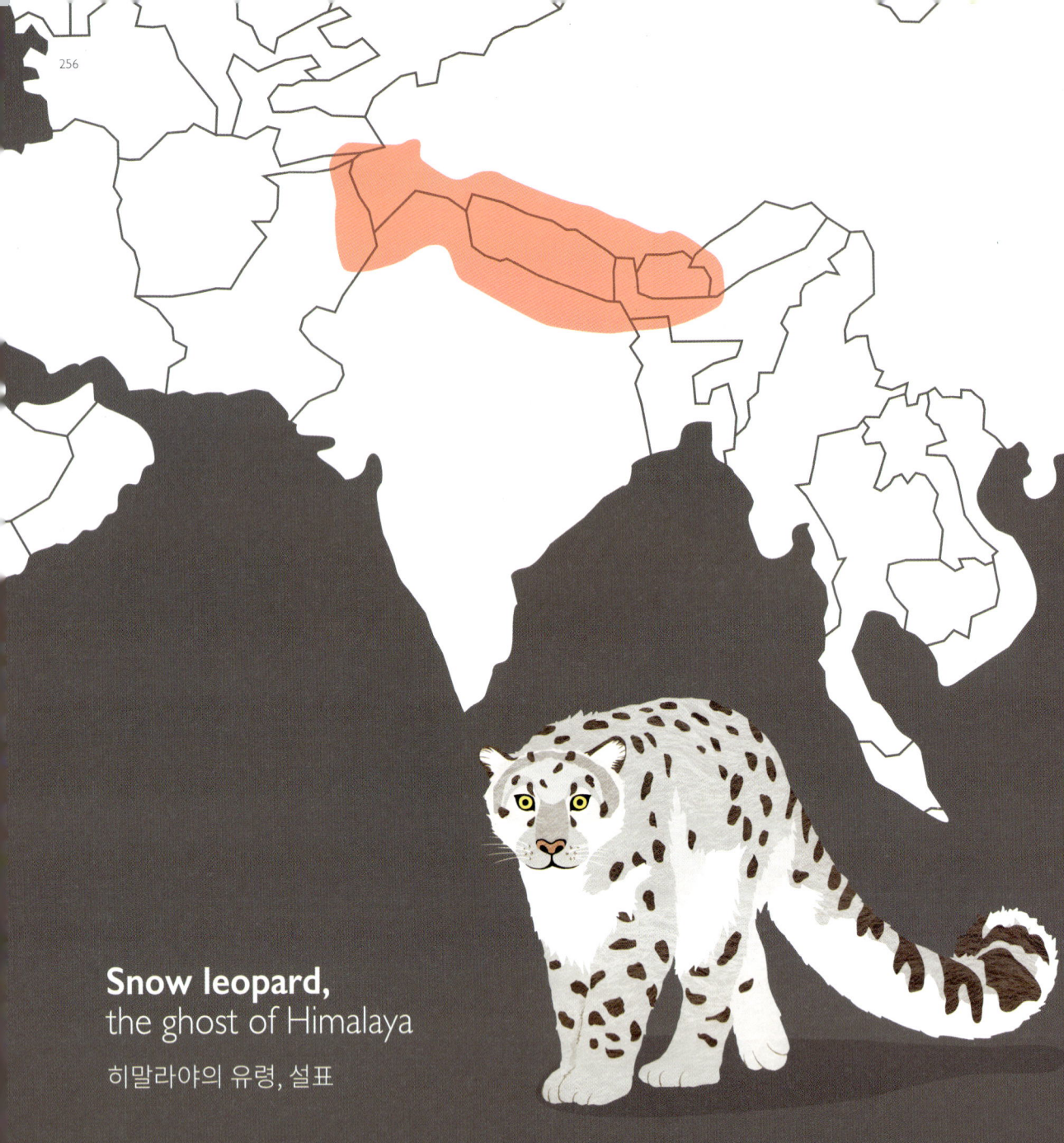

Snow leopard,
the ghost of Himalaya

히말라야의 유령, 설표

Size 크기	100~130cm / 25~75kg	Lifespan 평균수명	16 years in captivity 사육시 평균시 16년
Diet 식성	Carnivore 육식	Habitat 서식지	Mountain, savana, forest 산, 사바나, 숲
Behavior 행동	Solitary living 단독 생활	Distribution 분포지역	Central Asia 중앙 아시아
Reproduction 번식	Average 2 평균 2마리	Conservation status 멸종위기 등급	Vulnerable (VU) 취약종

EX Extinct
EW Extinct in the Wild
CR Critically Endangered
EN Endangered
VU Vulnerable

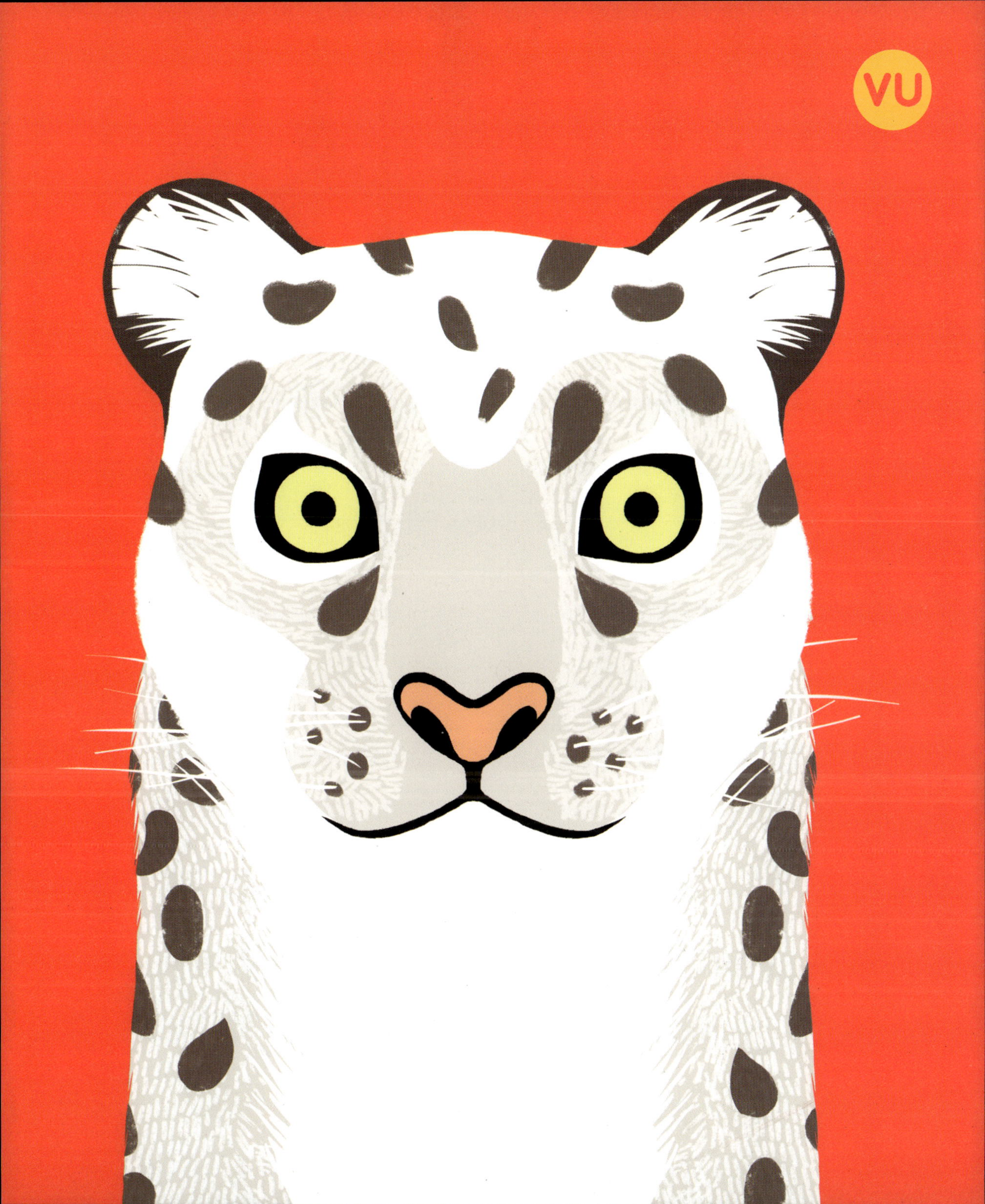

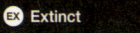

SNOW LEOPARD
VULNERABLE SPECIES

Slowloris,
clown on the tree
나무 위의 광대, 슬로우로리스

Size 크기	27~38cm / 0.6~0.7kg
Diet 식성	Omnivore 잡식
Behavior 행동	Solitary living 단독 생활
Reproduction 번식	Average 1 평균 1마리
Lifespan 평균수명	22 years
Habitat 서식지	Rainforest 열대우림
Distribution 분포지역	South-east Asia, Borneo 동남아시아, 보르네오 섬
Conservation status 멸종위기 등급	Vulnerable (VU) 취약종

They are named after the way a Dutch sailor first walked of Loris when he got off the boat and looked like a clown.
네델란드 선원이 배에서 로리스를 데리고 내릴 때 걷는 모습이 광대 같다고 해서 붙여진 이름입니다.

Siberian weasel, violent hunter
난폭한 사냥꾼, 족제비

Size 크기	26~38cm / 0.4~0.8kg	**Lifespan** 평균수명	6 years
Diet 식성	Carnivore 육식	**Habitat** 서식지	Tundra, forest, mountain 툰드라, 숲, 산
Behavior 행동	Solitary living 단독 생활	**Distribution** 분포지역	East Asia, Russia, Tivet 동아시아, 러시아, 티벳
Reproduction 번식	Average 5 평균 5마리	**Conservation status** 멸종위기 등급	Least concern (LC) 관심대상종

Weasel has a sporting nature, unlikely the cute appearance.
족제비는 귀여워 보이는 외모와 달리 사냥을 즐기는 성질을 가지고 있습니다.

Siberian tiger,
the lovebird of mammalia

잉꼬부부, 시베리아 호랑이

Size 크기	240~330cm / 100~360kg	Lifespan 평균수명	25 years
Diet 식성	Carnivore 육식	Habitat 서식지	Mountain 산
Behavior 행동	Solitary living 단독 생활	Distribution 분포지역	Russia, North Korea 러시아, 북한
Reproduction 번식	Average 1 평균 1마리	Conservation status 멸종위기 등급	Endangered (EN) 위기종

SIBERIAN TIGER
ENDANGERED SPECIES

EX Extinct
EW Extinct in the Wild
CR Critically Endangered
EN Endangered
VU Vulnerable
NT Near Threatened
LC Least Concern

WE'VE BEEN SAVED —
HUMPBACK
WHALE

혹등 고래는 적도와 북극해를 제외한 전 지역에 서식하며 주로 대서양이나 태평양 연안에서 발견됩니다. 고래 고기의 상업적 가치가 높아지자 국가 단위의 사냥이 이뤄졌고 남획의 규모는 점점 커져갔습니다. 16세기 후반부터 상업적 남획이 시작되면서 19~20 세기를 기점으로 개체 수가 급감하였으며 1970년 멸종위기종으로 분류되었습니다. 1946년 국제포경위원회(IWC)가 설립되어 다양한 규제를 제정하기 시작했고 1966년 혹등 고래는 포경금지 동물로 분류되었으며 당시 개체수는 5,000마리 정도에 불과했습니다. 1982년에는 모든 고래종에 대한 상업용 포경이 금지되었습니다. 이 외에도 IWC는 해양 쓰레기, 기후 변화에 따른 해양 변화 연구, 선박 충돌을 막기 위한 방법을 모색하며 관련 기관에 교육을 진행했습니다.

지속적인 규제와 캠페인을 통해 사람들의 인식이 바뀌어가면서 개체수는 증가하기 시작했습니다. 고래 사냥을 전통으로 여겼던 문화가 점차 사라지거나, 다방면의 규제를 통해 국가 단위의 사냥 계획을 무산시키는 등 많은 변화를 이뤄내고 있습니다.

미국 워싱턴대와 국립해양대기청(NOAA)의 연구진은 대서양에 서식하는 혹등 고래의 개체 수가 과거 남획으로 멸종위기에 처하기 전의 93% 수준인 25,000마리까지 회복되었다고 추정했습니다. 이에 따라 2018년을 기준으로 IUCN은 혹등 고래를 관심 대상종 (LC)으로 멸종위기 등급을 낮추었으며 84,000여 마리가 현존하는 것으로 추정하고 있습니다.

멸종위기로부터 벗어난 동물들 - 혹등고래

The humpback whale, found in all areas except for the equatorial and Arctic oceans, is primarily located off the coasts of the Atlantic and Pacific oceans. With the increase in commercial value of whale meat, national hunting efforts ensued, and the scale of exploitation continued to increase. Beginning in the late 16th century, commercial whaling began, leading to a sharp decline in population from the 19th to the 20th centuries, and was classified as an endangered species by 1970. The International Whaling Commission (IWC) was established in 1946, and various regulations were developed, with the humpback whale classified as a protected animal in 1966, with a population of only about 5,000 individuals at the time. In 1982, commercial whaling was banned for all whale species.

Additionally, the IWC has conducted research on ocean garbage and oceanic changes due to climate change, as well as explored methods to prevent ship collisions while providing education to related organizations. Through continuous regulations and campaigns, public perception has changed, leading to an increase in the population of humpback whales. Traditional cultures that viewed whaling as a tradition are gradually disappearing, and various regulations have thwarted national hunting plans, resulting in significant changes.

Researchers at the University of Washington and the National Oceanic and Atmospheric Administration (NOAA) estimate that the humpback whale population in the Atlantic has recovered to pre-exploitation levels, reaching approximately 25,000 individuals, which is 93% of the previous population. As a result, in 2018, the International Union for Conservation of Nature (IUCN) lowered the humpback whale's conservation status from endangered to a species of "least concern" (LC), estimating a current population of around 84,000 individuals.

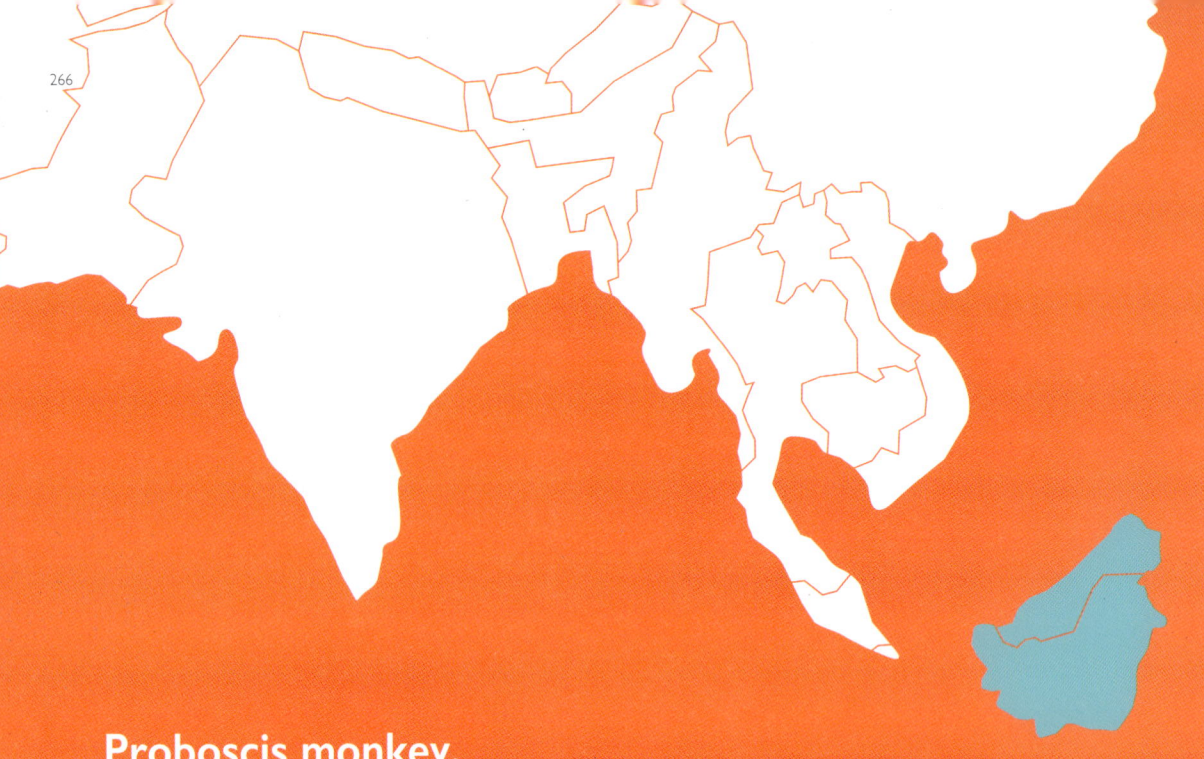

Proboscis monkey,
the most unique character
개성 넘치는 코주부 원숭이

Size 크기	60~70cm / 7~22kg
Diet 식성	Herbivore 초식
Behavior 행동	Group living 무리 생활
Reproduction 번식	Average 1 평균 1마리
Lifespan 평균수명	21 years in captivity 사육시 평균시 21년
Habitat 서식지	Mangrove, lowland rainforest 맹그로브, 저지대 열대우림
Distribution 분포지역	Borneo 보루네오 섬
Conservation status 멸종위기 등급	Endangered (EN) 위기종

PROBOSCIS MONKEY
ENDANGERED SPECIES

- EX Extinct
- EW Extinct in the Wild
- CR Critically Endangered
- **EN Endangered**
- VU Vulnerable
- NT Near Threatened
- LC Least Concern

Sugar Glider,
I'm not a pet

애완동물이 아니에요, 슈가글라이더

Size 크기	14~18cm / 0.1~0.15kg
Diet 식성	Omnivore 잡잡식
Behavior 행동	Group living 무리 생활
Reproduction 번식	Average 1~3 평균 1~3 마리
Lifespan 평균수명	9 years
Habitat 서식지	Forest, Savanna 숲, 사바나나
Distribution 분포지역	Australia, Indonesia, Papua New Guinea 호주, 인도네시아, 파푸아뉴기니
Conservation status 멸종위기 등급	Least concern (LC) 관심대상종

SUGAR GLIDER
LEAST CONCERN SPECIES

- **EX** Extinct
- **EW** Extinct in the Wild
- **CR** Critically Endangered
- **EN** Endangered
- **VU** Vulnerable
- **NT** Near Threatened
- **LC** Least Concern

Japanese marten,
golden suit gentleman
황금 옷의 신사, 산달

Size 크기	47~54.5cm / 0.5~1.7kg	**Lifespan** 평균수명	10 years
Diet 식성	Omnivore 잡식	**Habitat** 서식지	Coniferous forest 침엽수림
Behavior 행동	Solitary living 단독 생활	**Distribution** 분포지역	Japan 일본
Reproduction 번식	Average 1.5 평균 1.5마리	**Conservation status** 멸종위기 등급	Least concern (LC) 관심대상종

Japanese monkey,
hot spring mania
온천 매니아, 일본원숭이

Size 크기	52~57cm / 8~12kg	**Lifespan** 평균수명	25 years
Diet 식성	Omnivore 잡식	**Habitat** 서식지	Coniferous forest 침엽수림
Behavior 행동	Group living 무리생활	**Distribution** 분포지역	Japan 일본
Reproduction 번식	Average 1~2 평균 1~2마리	**Conservation status** 멸종위기 등급	Least concern (LC) 관심대상종

To withstand the cold, a Japanese monkey enjoys hot springs.
추위를 견디기 위해 온천을 즐깁니다.

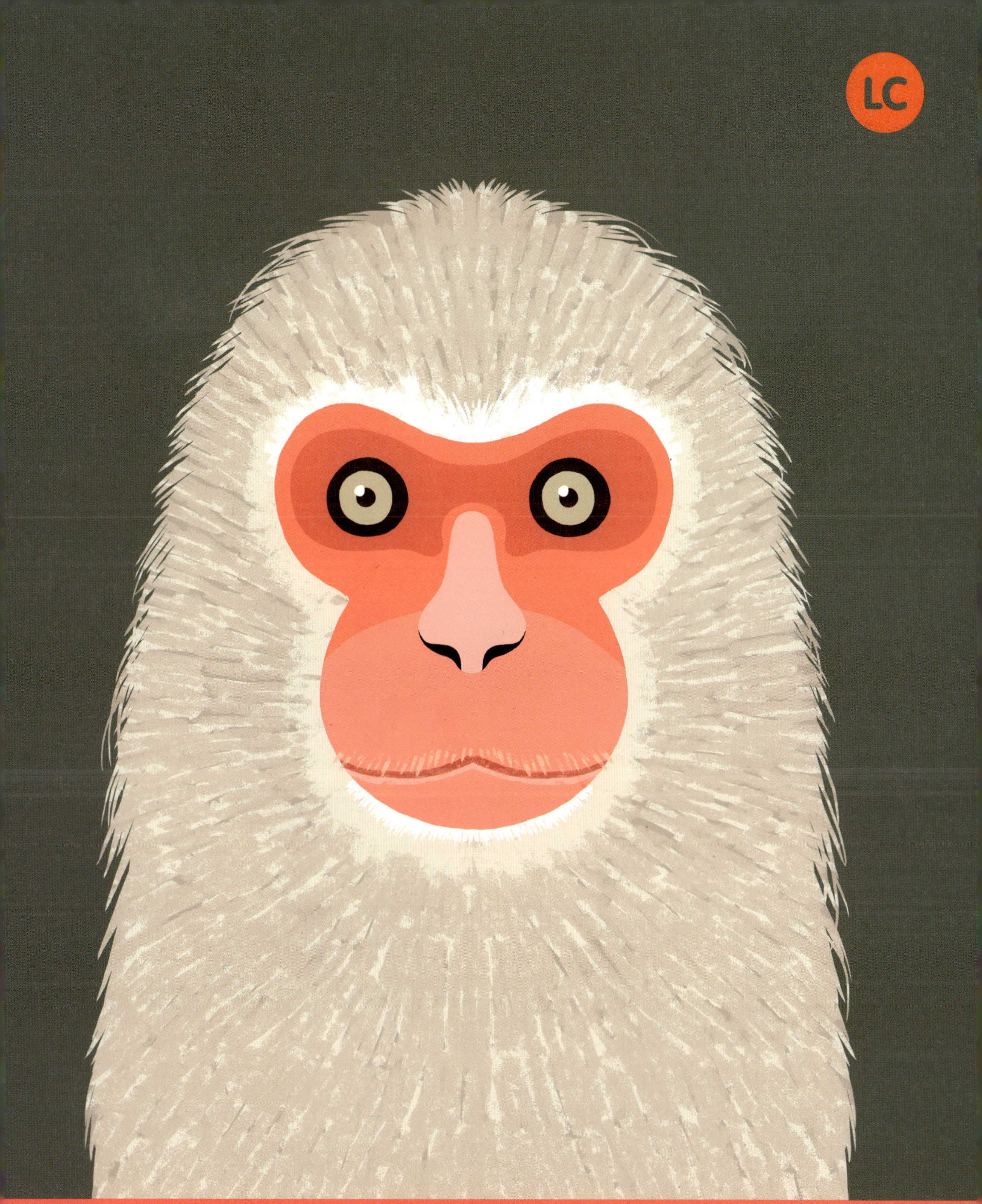

EX Extinct
EW Extinct in the Wild
CR Critically Endangered
EN Endangered
VU Vulnerable
NT Near Threatened
LC Least Concern

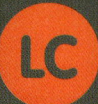

JAPANESE MONKEY
LEAST CONCERN SPECIES

Giant panda,
king of eating bamboo
대나무 푸드파이터, 자이언트 판다

Size 크기	150~180cm / 80~125kg	**Lifespan** 평균수명	12.5 years
Diet 식성	Herbivore 초식	**Habitat** 서식지	Forest 숲
Behavior 행동	Solitary living 단독 생활	**Distribution** 분포지역	Central part of China 중국 중앙 일부 지역
Reproduction 번식	Average 1.5 평균 1.5마리	**Conservation status** 멸종위기 등급	Vulnerable (VU) 취약종

Eurasian tawny owl,
Harry Potter's partner
해리포터의 파트너, 올빼미

Size 크기	38cm / 0.42~0.52kg	**Lifespan** 평균수명	15 years
Diet 식성	Carnivore 육식	**Habitat** 서식지	Forest, flatland 숲, 평지
Behavior 행동	Solitary living 단독 생활	**Distribution** 분포지역	Europe, Eurasia 유럽, 유라시아 지역
Reproduction 번식	Average 2 평균 2마리	**Conservation status** 멸종위기 등급	Least concern (LC) 관심대상종

LC

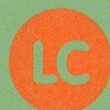

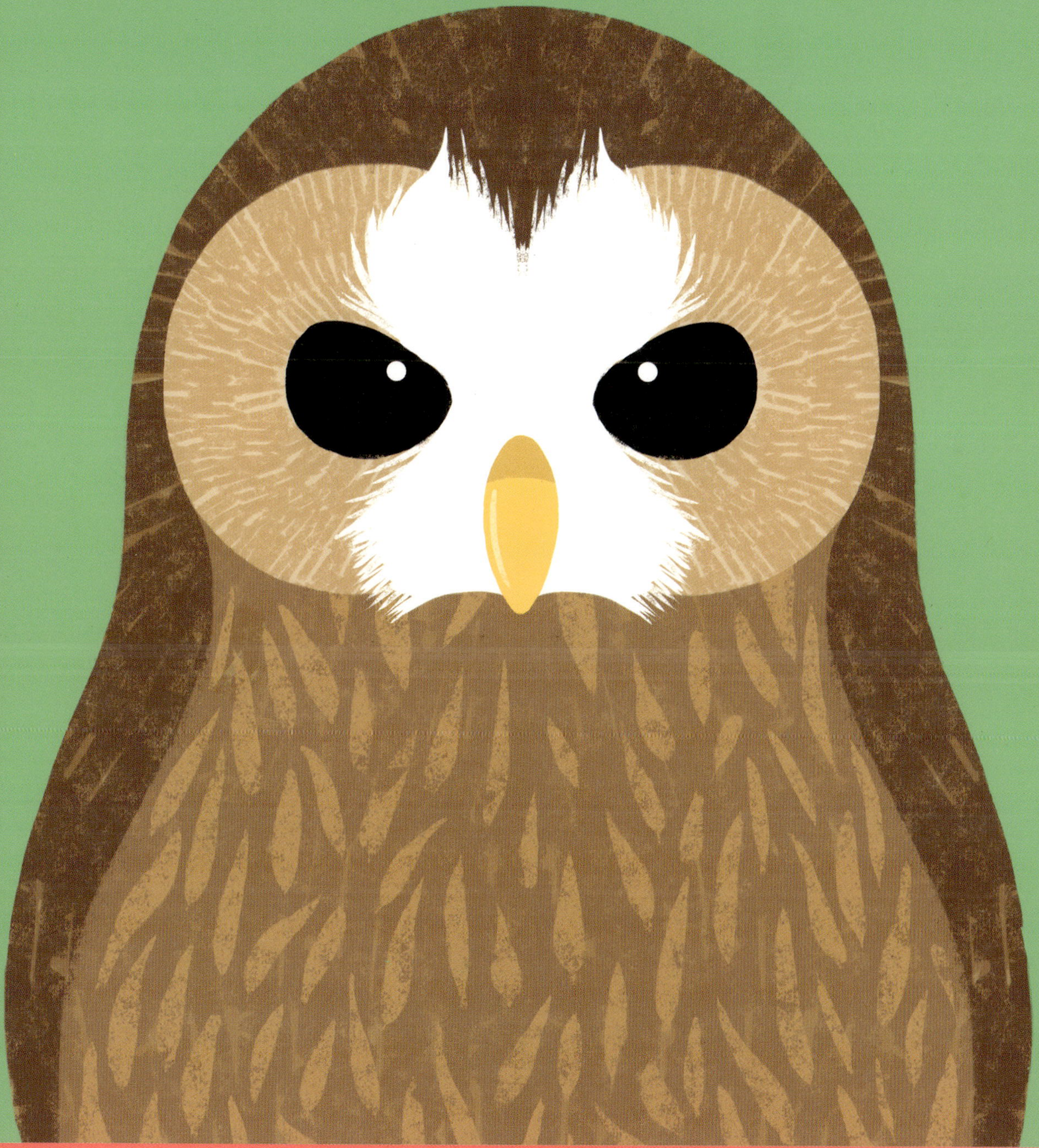

EURASIAN TAWNY OWL
LEAST CONCERN SPECIES

- EX Extinct
- EW Extinct in the Wild
- CR Critically Endangered
- EN Endangered
- VU Vulnerable
- NT Near Threatened
- LC Least Concern

Siau island tarsier,
inhabitant of tiny volcano island
작은 화산섬의 주민, 시아우섬 안경원숭이

Size 크기	9~16cm / 0.1~0.14kg	**Lifespan** 사육 평균수명	15 years in captivity 사육시 평균시 15년
Diet 식성	Omnivore 잡식	**Habitat** 서식지	Rainforest 열대우림
Behavior 행동	Group living 무리생활	**Distribution** 분포지역	Island of Siau 시아우 섬
Reproduction 번식	Average 1 평균 1마리	**Conservation status** 멸종위기 등급	Critically endangered (CR) 위급종

SIAU ISLAND TARSIER
CRITICALLY ENDANGERED SPECIES

EX Extinct
EW Extinct in the Wild
CR Critically Endangered
EN Endangered
VU Vulnerable
NT Near Threatened
LC Least Concern

WE'VE BEEN SAVED—KIHANSI SPRAY TOAD

탄자니아 동부 아크산맥만의 고유종으로, IUCN에 의해 야생에서 멸종된 것으로 분류되지만 포획 번식 프로그램을 통해 개체 수가 지속되고 있습니다. 야생에서 멸종되기 전(약 1999년)에는 약 17,000마리가 서식했던 것으로 추정됩니다. 하지만 2004년엔 암컷 3마리와 수컷 2마리만 발견될 정도로 급격히 감소하였다가, 결국 2009년 5월 야생에서 완전히 멸종되었다고 보고되었습니다. 원인은 키한시 스프레이 두꺼비는 폭포에서 협곡으로 떨어지며 발생하는 물로 습도를 유지하며 살아가는데 해당 지역에 댐이 건설되면서 서식지가 사라지게 되자 멸종하게 되었습니다.

탄자니아 정부는 2000년부터 댐 건설 시 서식 환경을 최대한 재현하기 위해 인공 폭포 시스템을 설치하여 키한시 스프레이 두꺼비의 서식지를 재조성 했습니다. 2012년 48마리를 야생으로 돌려보내기를 시작해, 번식 프로그램을 통한 지속적인 방생을 진행하고 있습니다.

멸종위기로부터 벗어난 동물들 - 키한시 스프레이 두꺼비

The Kihansi spray toad is a unique species found exclusively in the Eastern Arc Mountains of Tanzania. Despite being classified as extinct in the wild by the IUCN, the species continues to survive through a captive breeding program. It is estimated that approximately 17,000 individuals lived in the wild prior to their extinction in the late 1990s. However, their numbers sharply declined to only three females and two males by 2004, leading to their reported extinction in the wild in May 2009. The toad relies on the mist generated by waterfalls to maintain its habitat, and the construction of a dam in the area caused their habitat to disappear, ultimately leading to their extinction.

To revive the Kihansi spray toad population, the Tanzanian government reconstructed their habitat by installing an artificial waterfall system to recreate the mist generated by natural waterfalls. This system is crucial for the survival of the species. Since 2000, the government has been working to reproduce the toads in captivity and release them into the wild. In 2012, they released 48 individuals into the wild and have continued to breed the toads through a captive breeding program. The Tanzanian government remains dedicated to monitoring and protecting the species' habitat, as well as conducting research on their interactions with other species and fungi.

Lesser panda,
origin word of panda

팬더 단어의 시초, 랫서팬더

Size 크기	56~62.5cm / 4~7kg	**Lifespan** 평균수명	9 years
Diet 식성	Herbivore 초식	**Habitat** 서식지	Coniferous forest 침엽수림
Behavior 행동	Solitary living 단독 생활	**Distribution** 분포지역	Himalayas, Nepal, India 히말라야, 네팔, 인도
Reproduction 번식	Average 2 평균 2마리	**Conservation status** 멸종위기 등급	Endangered (EN) 위기종

Lesser panda was attached the name panda earlier than Giantpanda.
랫서팬더는 일반적으로 판다로 불리는 대왕판다 보다 더 먼저 판다라는 이름이 붙여졌습니다.

LESSER PANDA
ENDANGERED SPECIES

- **EX** Extinct
- **EW** Extinct in the Wild
- **CR** Critically Endangered
- **EN** Endangered
- **VU** Vulnerable
- **NT** Near Threatened
- **LC** Least Concern

Asian black bear,
the officer of forest
숲의 관리자, 반달가슴곰

Size 크기	120~180cm / 65~150kg	Lifespan 평균수명	25 years
Diet 식성	Omnivore 잡식	Habitat 서식지	Mountain, moist forest 산, 습한 숲
Behavior 행동	Solitary living 단독 생활	Distribution 분포지역	Pakistan, Afghanistan, Vietnam, China, Thailand, Korea, Japan 파키스탄, 아프가니스탄, 베트남, 중국, 태국, 한국, 일본
Reproduction 번식	Average 2 평균 2마리	Conservation status 멸종위기 등급	Vulnerable (VU) 취약종

When they have food, Asian black bears break twigs so that small plants can glow.
먹이 활동 과정에서 나뭇가지를 부러뜨려 작은 식물이 자랄 수 있게 합니다.

ASIAN BLACK BEAR
VULNERABLE SPECIES

EX Extinct
EW Extinct in the Wild
CR Critically Endangered
EN Endangered
VU Vulnerable
NT Near Threatened
LC Least Concern

Tonkin snub-nosed monkey,
attractive pink lips

매력적인 분홍 입술, 통킹들창코원숭이

Size 크기	51~65cm / 8~14kg	**Lifespan** 평균수명	20 years
Diet 식성	Herbivore 초식	**Habitat** 서식지	Subtropical area 아열대 지역
Behavior 행동	Group living 무리생활	**Distribution** 분포지역	Vietnam 베트남
Reproduction 번식	Average 1 평균 1마리	**Conservation status** 멸종위기 등급	Critically endangered (CR) 위급종

TONKIN SNUB-NOSED MONKEY
CRITICALLY ENDANGERED SPECIES

EX Extinct
EW Extinct in the Wild
CR Critically Endangered
EN Endangered
VU Vulnerable
NT Near Threatened
LC Least Concern

Japanese Serow,
whistling runner
휘파람을 불며 달리는, 일본 산양

Size 크기	70~80cm / 30~45kg	**Lifespan** 평균수명	20~21 years
Diet 식성	Herbivore 초식	**Habitat** 서식지	Forest, Grassland 숲, 초원
Behavior 행동	Group living 무리생활	**Distribution** 분포지역	Japen 일본
Reproduction 번식	Average 1 평균 1마리	**Conservation status** 멸종위기 등급	Least concern (LC) 관심대상종

When Japanese serows sense danger, they flee while emitting a whistling sound.
일본 산양은 위험을 감지하면 휘파람 소리와 함께 도망갑니다.

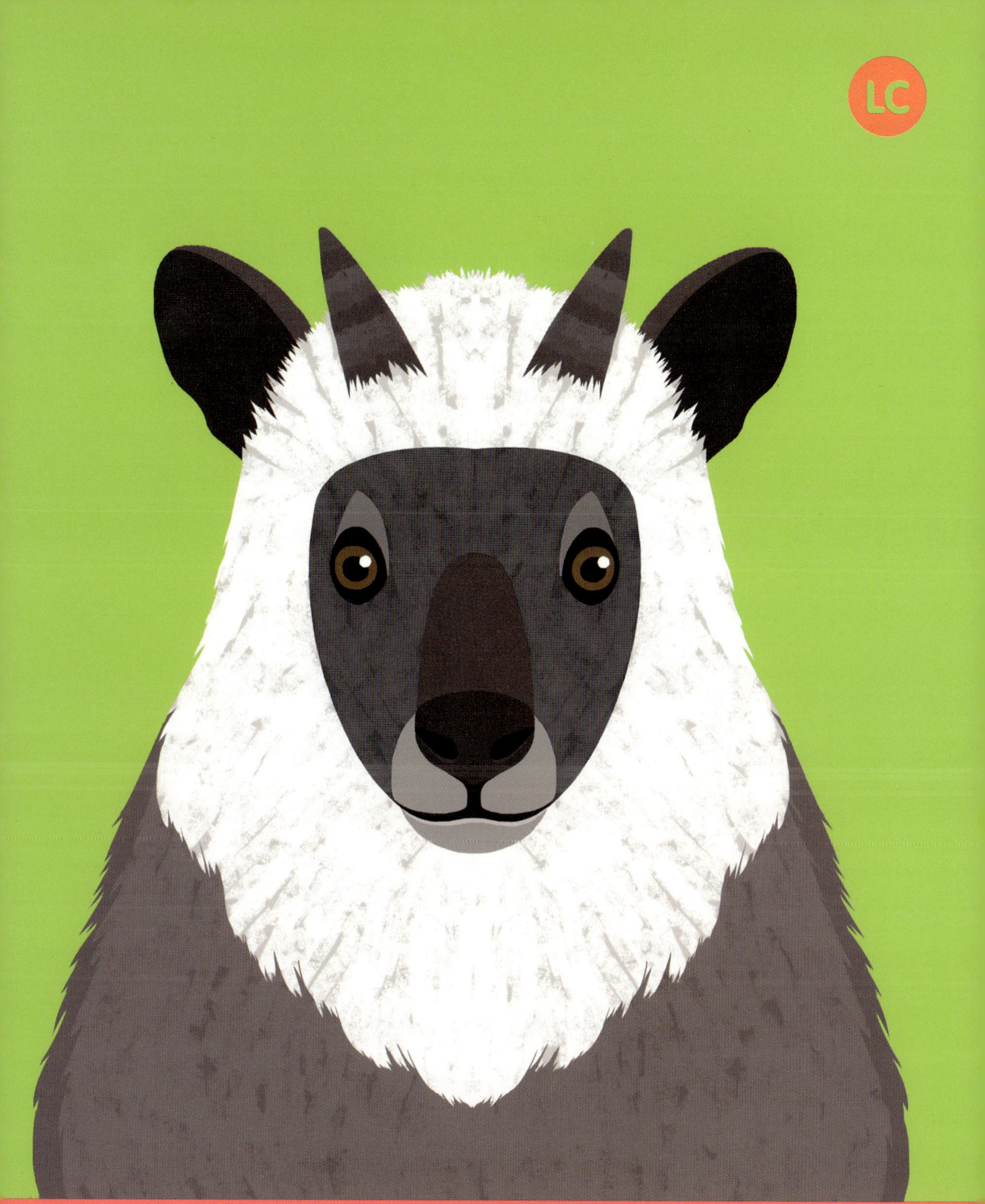

JAPANESE SEROW
LEAST CONCERN SPECIES

- EX Extinct
- EW Extinct in the Wild
- CR Critically Endangered
- EN Endangered
- VU Vulnerable
- NT Near Threatened
- LC Least Concern

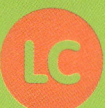

Matschie's tree kangaroo, tanning mania
태닝 매니아, 매치스나무캥거루

Size 크기	55~63cm / 6~13kg	**Lifespan** 평균수명	14 years	
Diet 식성	Herbivore 초식	**Habitat** 서식지	Lower montane forest 낮은 산지 숲	
Behavior 행동	Solitary living 단독 생활	**Distribution** 분포지역	Huon penisula of Papau New Guinea 파푸아뉴기니 후온반도	
Reproduction 번식	Average 1 평균 1마리	**Conservation status** 멸종위기 등급	Endangered (EN) 위기종	

A Matschie's tree kangaroo enjoys tanning in the trees.
매치스나무 캥거루는 나무 위에서 태닝하는 것을 즐깁니다.

Sun Bear,
world's smallest bear
세상에서 가장 작은 곰, 썬베어

Size 크기	120~150cm / 27~80kg	Lifespan 평균수명	15~30 year
Diet 식성	Omnivore 잡식	Habitat 서식지	Tropical forest 열대우림
Behavior 행동	Solitary living 단독생활	Distribution 분포지역	Southeast Asia 동남아시아
Reproduction 번식	Average 1~2 평균 1~2마리	Conservation status 멸종위기등급	Vulnerable (VU) 취약종

Malay civet,
cat of the jungle

밀림 속 고양이, 말레이 시벳

Size 크기	43~71cm / 1~5kg	**Lifespan** 평균수명	15~20 years
Diet 식성	Carnivore 육식	**Habitat** 서식지	Tropical rainforest 열대우림
Behavior 행동	Solitary living 단독생활	**Distribution** 분포지역	Malay Peninsula and the island of Sumatra 말레이시아 반도, 수마트라섬
Reproduction 번식	Average 2 평균 2마리	**Conservation status** 멸종위기등급	Least concern (LC) 관심대상종

WE'VE BEEN SAVED — MYANMAR STAR TORTOISE

미얀마(구 버마) 별 거북은 미얀마에서만 서식하는 종으로, 애완동물과 식료품, 의약품으로 인기가 많아져 1990년대 중반부터 개체 수가 급격히 감소했습니다. IUCN이 심각한 멸종 위기종으로 규정한 이후에도 많은 수요와 비싼 가격으로 인해 불법 사냥이 계속 되었습니다. Turtle Survival Alliance 에서 조사한 결과 2000년대 중반 야생에서 멸종된 것으로 보고되었습니다.

미얀마 정부와 Yadanabon Zoo 등 보호 단체들은 별 거북의 멸종을 막기위해 각종 규제와 종 보전 프로그램을 진행했습니다. 1994년부터 야생 동물 보호 보존법을 통해 불법 거래된 개체들을 압수해 보호하는 동시에 사육 및 번식을 시작했습니다. 2004년 부터 200마리를 시작으로 2017년까지 번식 프로그램을 통해 14,000마리 부화에 성공했고, 이 중 1,000여 마리가 야생으로 돌아갔습니다. 이후 야생에서 알을 부화시켜 살아갈 수 있도록 별 거북의 둥지를 옮기는 프로젝트를 진행하고 야생 동물 보호 구역을 별도로 지정해 밀렵꾼들로부터 안전한 서식지를 제공했습니다.

또한 정부와 단체들은 서식지 지역 주민들에게 미얀마 별거북에대한 문화적 공동체 의식을 키울 수 있도록 교육을 하는 등 종의 보호를 위해 다각적으로 방법을 모색하고 있습니다.

멸종위기로부터 벗어난 동물들 - 미얀마 별거북

The star tortoise of Myanmar (formerly known as Burma) is a species that only inhabits Myanmar, and it had become popular as a pet, food, and medicine, causing a sharp decline in its population since the mid-1990s. Despite being designated as a critically endangered species by the IUCN, illegal hunting had continued due to high demand and high prices. According to research by the Turtle Survival Alliance, it was reported in the mid-2000s that the species had become extinct in the wild.

In an effort to prevent the extinction of the star tortoise, the Myanmar government and conservation organizations such as the Yadanabon Zoo have implemented various regulations and conservation programs. Since 1994, they have seized illegally traded specimens and protected them while also starting breeding programs. From 2004 to 2017, starting with 200 specimens, they successfully hatched 14,000 tortoises, and more than 1,000 of them were released back into the wild. Subsequently, they also conducted a project to move star tortoise nests to hatch eggs and provide a safe habitat from poachers by designating protected areas for wildlife conservation.

In addition, the government and organizations are actively exploring various ways to protect the species, such as educating local communities about the cultural significance of the star tortoise in an effort to raise awareness and promote conservation efforts.

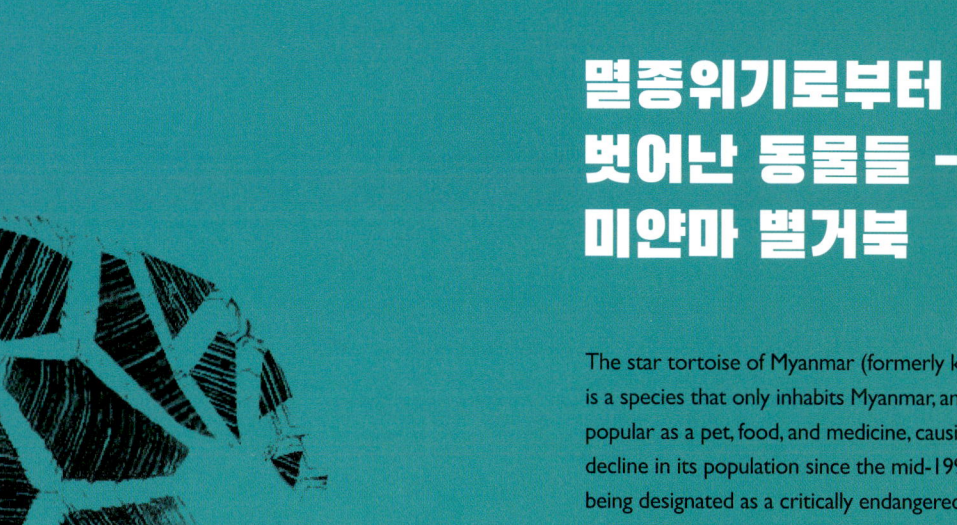

Indian giant squirrel, muscular tough guy
근육질 터프가이, 인도 큰 다람쥐

Size 크기	25~45cm / 1.5~3kg	**Lifespan** 평균수명	20 years in captivity 사육시 평균 20년
Diet 식성	Omnivore 잡식	**Habitat** 서식지	Forest 숲
Behavior 행동	Solitary living 단독생활	**Distribution** 분포지역	India 인도
Reproduction 번식	Average 1 평균 1마리	**Conservation status** 멸종위기등급	Least concern (LC) 관심대상종

Indian giant squirrel is 25~45cm long and has big strong claws to grab onto barks of trees and branches.
인도 큰 다람쥐는 몸길이 25~45cm에 나무껍질과 가지를 쥐기에 좋은 크고 강한 발톱을 가졌습니다.

Gaur,
bulky shy

수줍은 큰 덩치, 가우어

Size 크기	2.5~3.3m / 0.6~1t	Lifespan 평균수명	26 years in captivity 사육시 평균 26년
Diet 식성	Herbivore 초식	Habitat 서식지	Rainforest 열대우림
Behavior 행동	Group living 무리생활	Distribution 분포지역	Nepal, India, Myanmar 네팔, 인도, 미얀마
Reproduction 번식	Average 1 평균 1마리	Conservation status 멸종위기등급	Valunerable (VU) 취약종

Although Gaur has a big physique, it's timid and shy.
가우어는 거대한 체구와 힘에 비해 소심하고 수줍음이 많습니다.

Black javan leopard,
big pretty kitty

크고 아름다운 검은 고양이, 검은 자바 표범

Size 크기	90~150cm / 60~95kg	**Lifespan** 평균수명	21~23 years in captivity 사육시 평균 21~23년
Diet 식성	Carnivore 육식	**Habitat** 서식지	Rainforest 열대우림
Behavior 행동	Solitary living 단독생활	**Distribution** 분포지역	Indonesian island of Java 인도네시아 자바섬
Reproduction 번식	Average 4 평균 4마리	**Conservation status** 멸종위기등급	Endangered (EN) 위기종

Natuna island surili,
goggled monkey
뿔테안경을 낀, 나투나섬잎 원숭이

Size 크기	42~61cm / 5~9kg	**Lifespan** 평균수명	15~25 years in captivity 사육시 15~25년
Diet 식성	Herbivore 초식	**Habitat** 서식지	Forest 숲
Behavior 행동	Group living 무리 생활	**Distribution** 분포지역	Indonesian island of Natuna Besar 인도네시아 나투나제도
Reproduction 번식	Average 1 평균 1마리	**Conservation status** 멸종위기등급	Vulnerable (VU) 취약종

NATUNA ISLAND SURILI
VULNERABLE SPECIES

EX Extinct
EW Extinct in the Wild
CR Critically Endangered
EN Endangered
VU Vulnerable
NT Near Threatened
LC Least Concern

Sumatran orangutan, forest man
숲의 사람, 수마트라 오랑우탄

Size 크기	90~140cm / 45~90kg	Lifespan 평균수명	44~58 years
Diet 식성	Herbivore 초식	Habitat 서식지	Rainforest 열대우림
Behavior 행동	Group living 무리생활	Distribution 분포지역	Island of Sumatra in Indonesia 인도네시아 수마트라섬
Reproduction 번식	Average 1 평균 1마리	Conservation status 멸종위기등급	Critically endangered (CR) 위급종

'Orangutan' means 'person of the forest' in local Malaysian.
'오랑우탄'이라는 이름은 말레이시아 원주민어로 '숲의 사람'이라는 뜻입니다.

SUMATRAN ORANGUTAN
CRITICALLY ENDANGERED SPECIES

- EX Extinct
- EW Extinct in the Wild
- CR Critically Endangered
- EN Endangered
- VU Vulnerable
- NT Near Threatened
- LC Least Concern

Indian muntjac, roaring deer
짖는 사슴, 인도 문착

Size 크기	150~170cm / 60~125kg	**Lifespan** 평균수명	25~29 years
Diet 식성	Herbivore 초식	**Habitat** 서식지	Desert, steppe 사막, 반건조 초원
Behavior 행동	Group living 무리생활	**Distribution** 분포지역	Southeast Asia 동남아시아
Reproduction 번식	Average 1 평균 1마리	**Conservation status** 멸종위기등급	Least concern (LC) 관심대상종

Indian Muntiacini are known as 'roaring deer' because roar when they face their predators to alert others of dangers.
인도 문착은 포식자와 마주쳤을 때 울부짖는 듯한 소리로 위험함을 알려 '짖는 사슴'이라고도 불립니다.

Ili pika,
the model for Pikachu
피카츄의 모델, 일리 우는토끼

Size 크기	20~23cm / 0.2~0.25kg	**Lifespan** 평균수명	3~7 years
Diet 식성	Herbivore 초식	**Habitat** 서식지	High cliff 높은 절벽
Behavior 행동	Group living 무리생활	**Distribution** 분포지역	Northwest China 중국 북서쪽 지역
Reproduction 번식	Average 1~2 평균 1~2마리	**Conservation status** 멸종위기등급	Endangered (EN) 위기종

ILI PIKA
ENDANGERED SPECIES

EX Extinct
EW Extinct in the Wild
CR Critically Endangered
EN Endangered
VU Vulnerable
NT Near Threatened
LC Least Concern

WE'RE BEING MISTREATED— WATER DEER

안녕하세요, 저는 고라니입니다.
세계 자연보전 연맹은 저를 멸종 위기 등급 '취약종'으로 지정하고 있지만, 대한민국에서는 '야생 유해 동물'로 부릅니다. 또 다른 서식지인 중국에서는 2급 국가 보호 동물로 지정해 보호하고 있는데 전 세계 개체 수 90%가 서식하는 대한민국에서 저는 그저 애물단지에 불과합니다.
저는 왜 유해 동물이 되었을까요? 어린잎을 좋아하고, 그저 살던 곳의 길을 지나갔을 뿐인데 사람들은 말합니다. 애써 가꾼 농작물을 다 자라기도 전에 망가뜨리고, 갑자기 차도에 튀어나와 자동차를 놀라게 하기 때문이라고 말입니다. 환경부가 2020년 발표한 자료에 따르면 최근 5년간 발생한 전국 야생 동물 찻길 사고의 60%가 저 때문에 일어난 사고였으며, 충북에서 긴급 구조된 제 친구들의 80%가 교통사고에 의한 부상이었습니다.

이처럼 이곳에서는 저와 친구들을 흔하게 볼 수 있고, 인간에게 직접적인 피해를 준다는 인식 때문에 보호해야 할 멸종 위기의 동물이라고 생각하기 어렵습니다.

Hi there, I'm a Gorani. a.k.a. Korean water deer. While the International Union for Conservation of Nature (IUCN) designates me as a vulnerable species, in South Korea, I'm referred to as a "wildlife pest". In China, another habitat of mine, I'm classified as a second-class protected animal, but in South Korea, where 90% of my population lives, I'm just a nuisance.
I wonder why I'm considered a harmful animal? I simply enjoy eating young leaves and passing through the places where I live. But people say that I cause harm because I often damage crops that humans have carefully grown, and I sometimes dart onto roads and startle drivers. According to a report by the South Korean Ministry of Environment released in 2020, 60% of the country's wildlife accidents on roads over the past five years were caused by me, and 80% of my friends who were urgently rescued in Chung-cheong bukdo province were injured due to traffic accidents.

As a result, it's difficult to think of me and my friends as endangered species that need protection because we are often seen as a direct threat to humans in South Korea.

우리는 억울하다 ― 고라니

이 모든 일들이 다 저 때문일까요?

대한민국에서는 저를 잡아먹고 살아야 하는 호랑이나 곰과 같은 육식동물이 먼저 멸종 위기를 겪으며 야생에서 사라지자 저와 친구들의 수는 많아졌지만, 살아가야 할 곳들은 도시 개발로 인해 점점 좁아지고 있다 보니 이런 사고들이 자꾸만 일어나는 것은 아닐까요? 이 외에도 최근 들어 기후 변화로 인해 자주 발생하는 산불은 우리가 사는 곳을 순식간에 잿더미로 만들어 버리기도 하지요. 그럼 또 먹을 것이 없어진 우리는 어린잎을 찾아 마을을 헤매야 하고, 길을 건너다 자동차에 치여 죽게 되겠지요.

과연 저는 해로운 동물일까요?

Are all of these things my fault?

As carnivorous animals like tigers and bears, who have to hunt and eat me to survive, have faced extinction in Korea, the population of me and my friends has increased. However, as living spaces are getting smaller and smaller due to urban development, aren't these accidents happening more frequently? In addition, frequent wildfires caused by climate change have also turned our homes into ashes in no time. Then we will have nothing to eat and have to wander around the village looking for food, only to be hit by a car while crossing the road.

Am I really a harmful animal?

Javan hawk-eagle,
iconic hairstyle

독보적인 헤어스타일, 자바 뿔매

Size 크기	56~60cm	**Habitat** 서식지	Tropical forest 열대우림
Diet 식성	Carnivore 육식	**Distribution** 분포지역	Island of Java in Indonesia 인도네시아 자바섬
Behavior 행동	Solitary living 단독생활	**Conservation status** 멸종위기등급	Endangered (EN) 위기종
Reproduction 번식	Average 1 egg 평균 1개의 알		

Water Deer,
we all live together on the Korean Peninsula.
한반도에 모여 살아요, 고라니

Size 크기	77~100cm / 9-14kg	**Lifespan** 평균수명	6 years
Diet 식성	Herbivore 초식	**Habitat** 서식지	Forest, Grassland 숲, 초원
Behavior 행동	Solitary living 단독생활	**Distribution** 분포지역	China, Korea 중국, 한국
Reproduction 번식	Average 2~3 평균 2~3마리	**Conservation status** 멸종위기등급	Vulnerable (VU) 취약종

WATER DEER
VULNERABLE SPECIES

EX Extinct
EW Extinct in the Wild
CR Critically Endangered
EN Endangered
VU Vulnerable
NT Near Threatened
LC Least Concern

Plateau Pika,
diligent all year round
일년내내 부지런한, 고원 우는토끼

Size 크기	14~18cm / 0.1~0.2kg	**Lifespan** 평균수명	16 years
Diet 식성	Herbivore 초식	**Habitat** 서식지	Grassland, Desert 초원, 사막
Behavior 행동	Group living 무리생활	**Distribution** 분포지역	China, India, Nepal 중국, 인도, 네팔
Reproduction 번식	Average 2~5 평균 2~5마리	**Conservation status** 멸종위기등급	Least concern (LC) 관심대상종

Although Plateu Pika lives in an extremely cold environment, it does not hibernate during the winter.
고원 우는토끼는 극도로 추운 환경에서 살고 있지만 겨울잠을 자지 않습니다.

PLATEAU PIKA
LEAST CONCERN SPECIES

- **EX** Extinct
- **EW** Extinct in the Wild
- **CR** Critically Endangered
- **EN** Endangered
- **VU** Vulnerable
- **NT** Near Threatened
- **LC** Least Concern

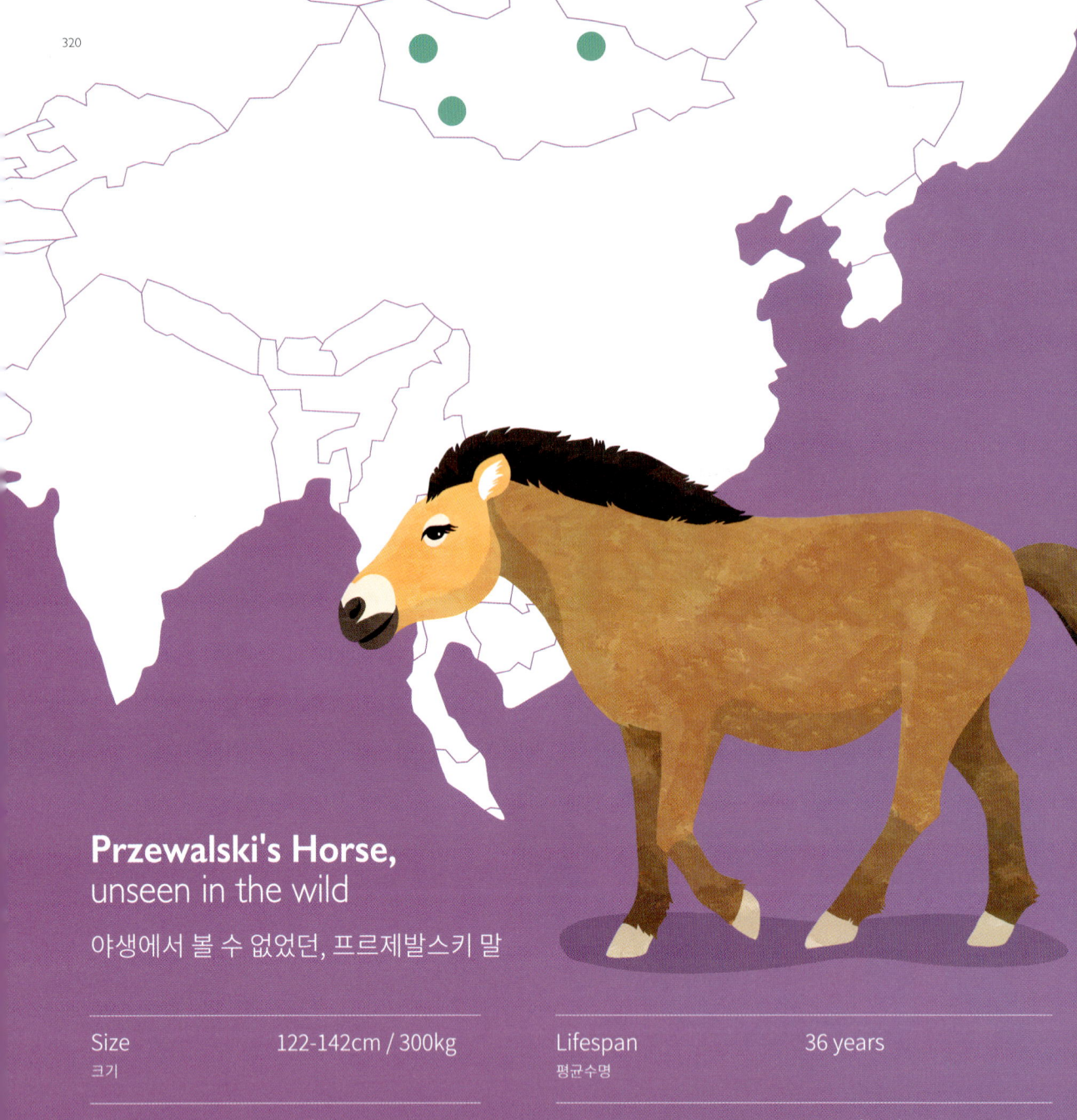

Przewalski's Horse, unseen in the wild

야생에서 볼 수 없었던, 프르제발스키 말

Size 크기	122-142cm / 300kg	**Lifespan** 평균수명	36 years
Diet 식성	Herbivore 초식	**Habitat** 서식지	Grassland, Desert 초원, 사막
Behavior 행동	Group living 무리생활	**Distribution** 분포지역	Mongolia, Kazakhstan, Russian, Ukraine 몽골, 카자흐스탄, 러시아, 우크라이나
Reproduction 번식	Average 1 평균 1마리	**Conservation status** 멸종위기등급	Endangered (EN) 위기종

Przewalski's horse, extinct in the wild since the mid-20th century, has been preserved and reintroduced to the wild since the 1990s in Mongolia, Central Asia, and Eastern Europe through breeding programs.
야생에서 멸종된 후 1990년대부터 몽골, 중앙아시아, 동유럽 등에 보존 사육하여 야생으로 방생하였습니다.

Binturong, a fig mania
무화과 마니아, 빈투롱

Size 크기	60~97cm / 9~20kg	**Lifespan** 평균수명	18~20 years
Diet 식성	Omnivore 잡식	**Habitat** 서식지	Forest 숲
Behavior 행동	Solitary living 단독생활	**Distribution** 분포지역	India, Nepal, Bangladesh, Bhutan, Myanmar 인도, 네팔, 방글라데시, 부탄, 미얀마
Reproduction 번식	Average 1~6 평균 1~6마리	**Conservation status** 멸종위기등급	Vulnerable (VU) 취약종

EX Extinct
EW Extinct in the Wild
CR Critically Endangered
EN Endangered
VU Vulnerable
NT Near Threatened
LC Least Concern

BINTURONG
VULNERABLE SPECIES

Yellow-eyed Penguin,
oldest living penguin species

현존하는 가장 오래된 펭귄종, 노란눈 펭귄

Size 크기	65~79cm / 5~8kg	Lifespan 평균수명	20 years
Diet 식성	Fish, Cephalopoda 생선, 두족류	Habitat 서식지	Forest, Marine Neritic 숲, 얕은 바다
Behavior 행동	Group living 무리생활	Distribution 분포지역	South-eastern New Zealand 뉴질랜드 남동부
Reproduction 번식	Average 2eggs 평균 2개의 알	Conservation status 멸종위기등급	Endangered (EN) 위기종

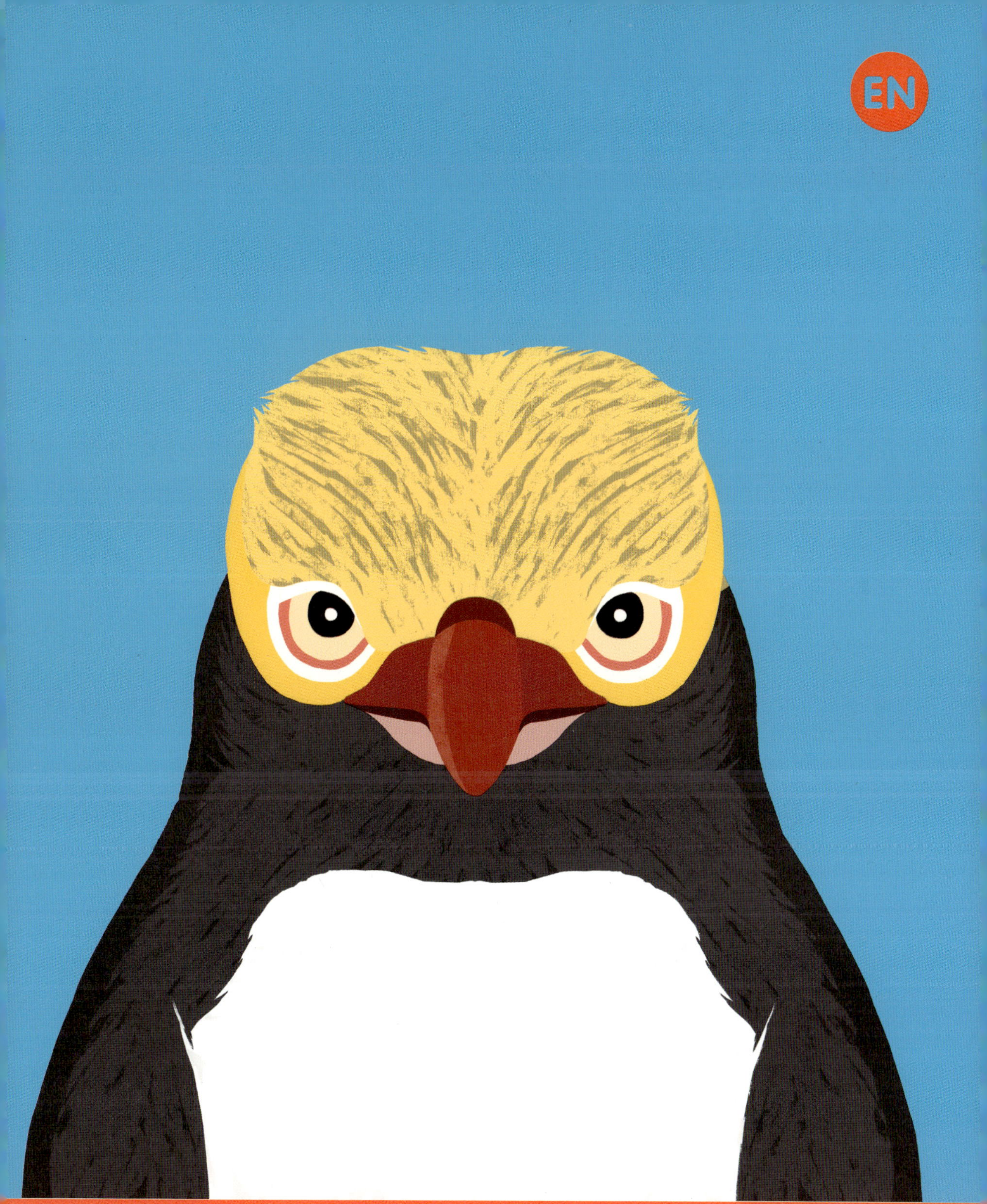

- EX Extinct
- EW Extinct in the Wild
- CR Critically Endangered
- EN Endangered
- VU Vulnerable
- NT Near Threatened
- LC Least Concern

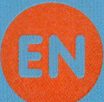

YELLOW-EYED PENGUIN
ENDANGERED SPECIES

WE'RE BEING MISTREATED— WILD BOAR

안녕하세요, 저는 야생 멧돼지입니다. 야생 멧돼지와 가축 돼지는 생김새가 다르지만, 사실은 같은 종인 친구입니다. 혹시 아프리카 돼지 열병을 알고 계시나요? 치명적 바이러스 전염병으로 감염률이 높고 치사율이 99%에 가까워 양돈 산업에 큰 피해를 주는 질병이지요. 백신이나 치료제가 없어 방역과 살처분 외에는 해결 방법이 없는 상황입니다. 1920년대 아프리카에서 처음 발생하여 현재까지 전 세계적으로 전파되었어요. 대한민국의 경우에는 2019년 9월 17일 경기 파주의 한 농장에서 처음 발생한 후 점차 남쪽으로 번지고 있습니다.

바이러스를 가지고 있는 제 친구가 북한에서 남한으로 넘어와 열병을 최초 전파했다고 알려져 있지만 이것은 정확한 사실이 아니에요. 2019년 5월 북한에서 처음으로 발생했지만 이후 얼마나 퍼졌는지 확인할 수 없으며, 멧돼지가 비무장 지대를 가로질러 남한으로 넘어와 가축 돼지 친구들에게 전파했을 가능성은 매우 낮아요. 애초에 아프리카에서 세계로 전파된 것은 바이러스에 접촉한 인간이 비행기나 자동차 같은 각종 이동 수단을 통해 대륙을 넘어 가축 돼지와 접촉했기 때문일 가능성이 더 큽니다.

Hello, I am a wild boar. Although wild boars & domestic pigs have different appearances, we're actually the same species and friends. Have you heard of African swine fever? It's a highly infectious and deadly viral disease with a mortality rate of nearly 99%, which causes significant damage to the swine industry as there is no cure or vaccine, except for preventative measures like culling and quarantine. It first emerged in Africa in the 1920s and has since spread worldwide. In South Korea, it was first reported on September 17, 2019, on a farm in Paju, Gyeonggi Province, and has been spreading gradually southward since then.

It has been rumored that my friend carrying the virus crossed over from North Korea to South Korea and was the first to spread the disease, but this isn't entirely accurate. Although African swine fever did occur in N. Korea in May 2019, it is unclear how far it spread, and the possibility of wild boars carrying the virus across the DMZ and infecting domestic pigs is low. In fact, it's more likely that the virus was transmitted from infected humans who had contact with pigs while traveling via various means of transportation such as airplanes or cars, crossing continents.

우리는 억울하다__
야생 멧돼지

저와 제 친구들도 물론 열병에 감염되긴 하지만 야생에서 자리 유전자가 다양하기 때문에 모든 개체가 감염되거나 죽지는 않아요. 하지만 가축 돼지 친구들의 치사율이 99%인 이유는 사람들이 오랜 세월 동안 맛이 좋고 빠르게 성장하는 돼지를 만들기 위해 유전자를 조작하여 다양성을 없애고 공장식으로 사육하기 때문이에요. 야생에서는 각 개체마다 면역력과 건강 상태가 다르므로 큰 문제가 되지 않는 병이 단일 유전자에 공장식으로 길러지는 가축들에게는 한 마리만 바이러스에 감염되어도 치명적일 수밖에 없습니다.

과연 제가 병균을 옮기는 유해한 동물일까요?

My friends & I can also contract the disease, but since we grow up in the wild, we have more genetic diversity and individual immunity, so not all of us get infected or die. However, the reason why domestic pigs have a mortality rate of 99% is because humans have been manipulating their genes to create pigs that taste good and grow fast, thereby reducing genetic diversity and mass-producing them in factories. In the wild, each individual has different immune responses and health conditions, so diseases are not a big problem. But for factory-raised pigs with identical genes, even one infected animal can cause devastating consequences.

So, am I really a harmful animal that spreads pathogens?

Sulawesi Serpent-eagle,
the guardian of the Sulawesi

섬의 수호자, 술라웨시 뱀독수리

Size 크기	46~54cm	**Lifespan** 평균수명	13 years	
Diet 식성	Carnivore 육식	**Habitat** 서식지	Forest, Savanna 숲, 사바나	
Behavior 행동	Solitary living 단독생활	**Distribution** 분포지역	Indonesia 인도네시아	
Reproduction 번식	Average 1~3 eggs 평균 1~3개의 알	**Conservation status** 멸종위기등급	Least concern (LC) 관심대상종	

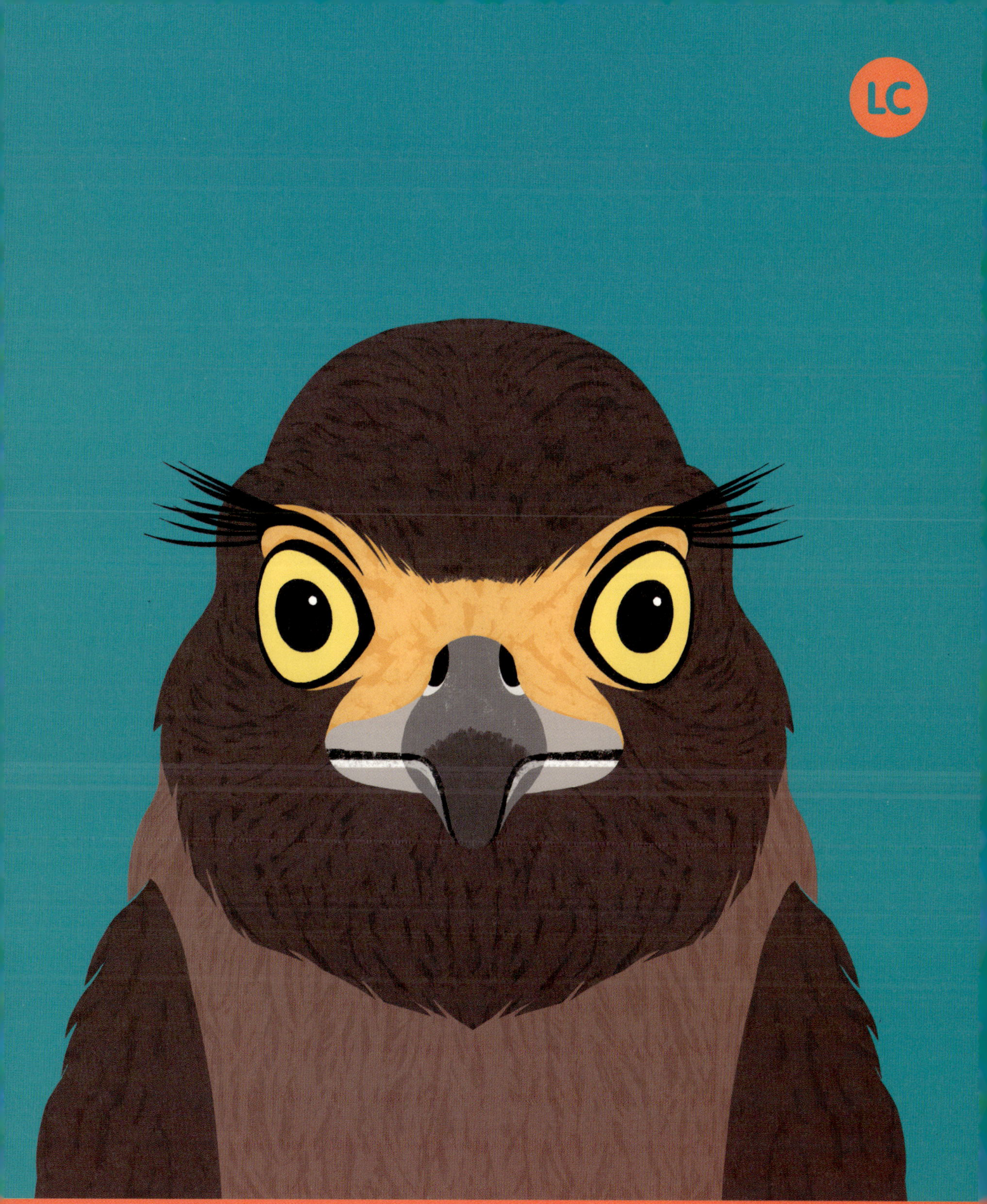

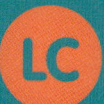

SULAWESI SERPENT-EAGLE
LEAST CONCERN SPECIES

- EX Extinct
- EW Extinct in the Wild
- CR Critically Endangered
- EN Endangered
- VU Vulnerable
- NT Near Threatened
- LC Least Concern

Manul,
the master of facial expressions

표정연기 마스터, 마눌

Size 크기	46-65cm / 2.5-4.5kg	**Lifespan** 평균수명	3.5 years
Diet 식성	Carnivore 육식	**Habitat** 서식지	Grassland, Rocky areas 초원, 바위 지역
Behavior 행동	Solitary living 단독생활	**Distribution** 분포지역	Afghanistan, Bhutan, China 아프가니스탄, 부탄, 중국
Reproduction 번식	Average 3~6 평균 3~6마리	**Conservation status** 멸종위기등급	Least concern (LC) 관심대상종

MANUL
LEAST CONCERN SPECIES

- **EX** Extinct
- **EW** Extinct in the Wild
- **CR** Critically Endangered
- **EN** Endangered
- **VU** Vulnerable
- **NT** Near Threatened
- **LC** Least Concern

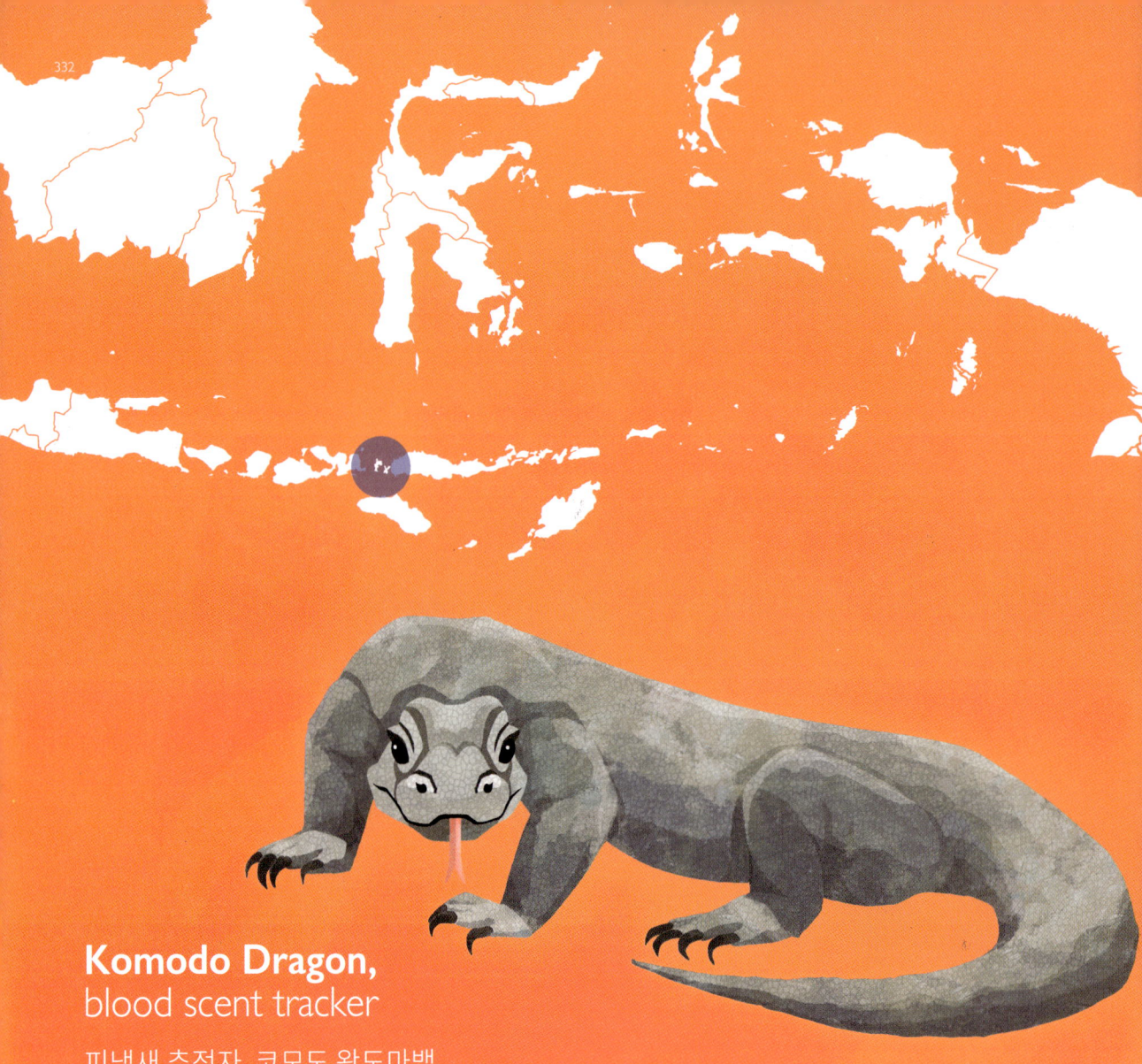

Komodo Dragon,
blood scent tracker
피냄새 추적자, 코모도 왕도마뱀

Size 크기	Up to 3m / 165kg 최대 크기	**Lifespan** 평균수명	40~50 years
Diet 식성	Carnivore 육식	**Habitat** 서식지	Forest, Savanna 숲, 사바나
Behavior 행동	Solitary living 단독생활	**Distribution** 분포지역	Komodo Island in southern Indonesia 인도네시아 남부 코모도섬
Reproduction 번식	Average 20eggs 평균 20개의 알	**Conservation status** 멸종위기등급	Endangered (EN) 위기종

Komodo dragon has a keen sense of smell and is sensitive to the scent of blood, capable of detecting it from a distance of up to 10 km away.
후각이 뛰어나고 피 냄새에 민감해 10km나 떨어져 있는 피냄새도 맡을 수 있습니다.

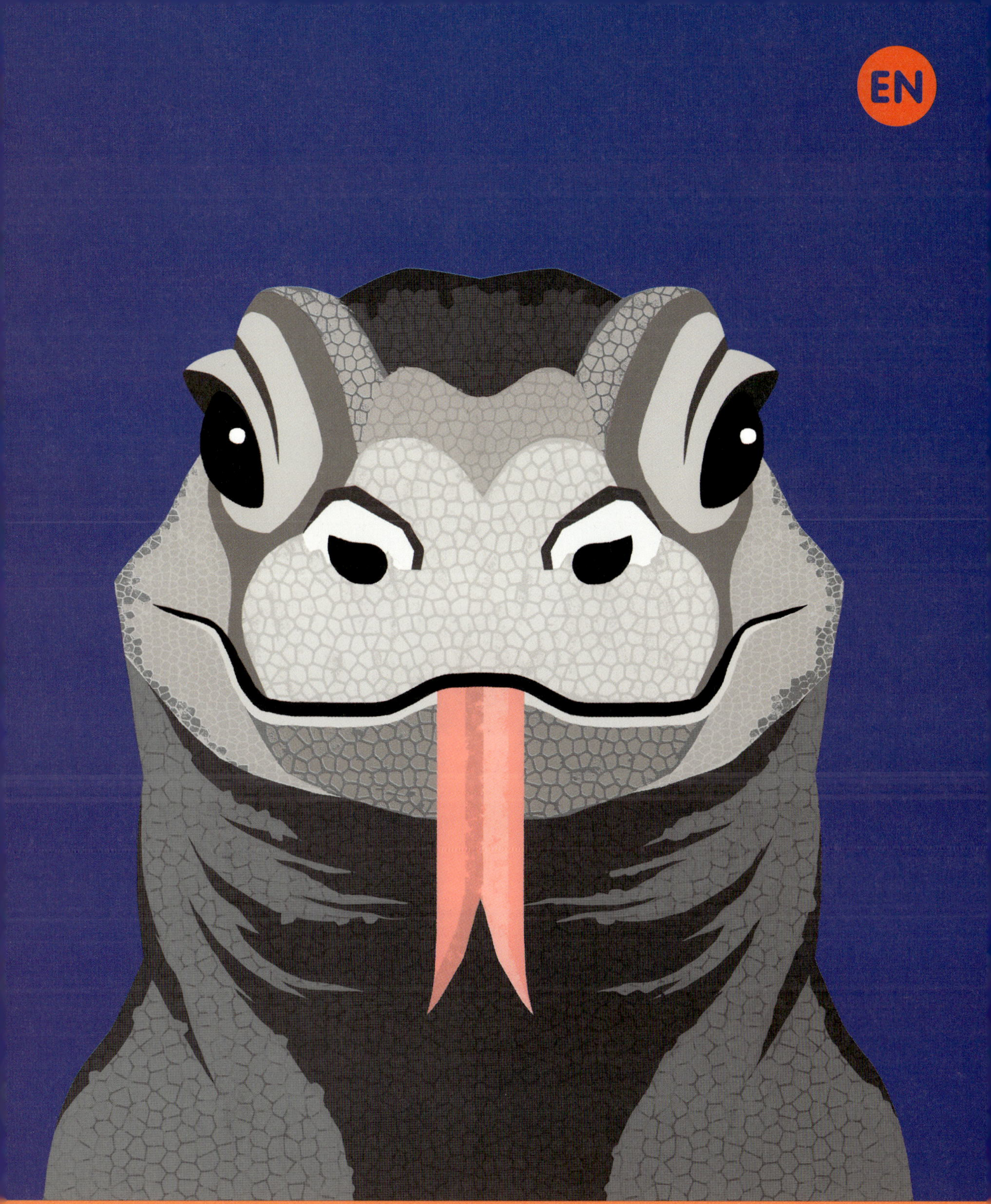

- EX Extinct
- EW Extinct in the Wild
- CR Critically Endangered
- EN Endangered
- VU Vulnerable
- NT Near Threatened
- LC Least Concern

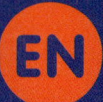

KOMODO DRAGON
ENDANGERED SPECIES

Flat-Headed Cat,
fisherman of the forest
숲 속의 낚시꾼, 납작머리 살쾡이

Size 크기	30~50cm / 1.5~2.75kg	**Lifespan** 평균수명	14years in captivity 사육시 14년
Diet 식성	Carnivore 육식	**Habitat** 서식지	Forest, Wetland 숲, 습지
Behavior 행동	Solitary living 단독생활	**Distribution** 분포지역	Malaysia, Sumatra, Borneo 말레이시아, 수마트라, 보르네오
Reproduction 번식	Average 1~4 평균 1~4마리	**Conservation status** 멸종위기등급	Endangered (EN) 위기종

Flat-Headed cat have excellent fishing skills thanks to their strong jaw strength, which allows them to catch fish with ease.
강력한 턱 힘 덕분에 물고기 잡는 실력이 뛰어납니다.

Golden snub-nosed monkey, golden alien
황금 외계인, 황금들창코원숭이

Size 크기	57~76cm / 15~30kg	Lifespan 평균수명	26 years in captivity 사육시 26년
Diet 식성	Herbivore 초식	Habitat 서식지	Forest, mountain 숲, 산
Behavior 행동	Group living 무리 생활	Distribution 분포지역	Sichuan, Gansu, Shanxi of China 중국 쓰촨, 간수, 산시
Reproduction 번식	Average 1 평균 1마리	Conservation status 멸종위기 등급	Endangered (EN) 위기종

GOLDEN SNUB-NOSED MONKEY
ENDANGERED SPECIES

- EX Extinct
- EW Extinct in the Wild
- CR Critically Endangered
- EN Endangered
- VU Vulnerable
- NT Near Threatened
- LC Least Concern

Prevost's Squirrel,
dressed in an orange suit

오렌지 수트를 입은, 삼색다람쥐

Size 크기	20~27cm / 0.25~0.5kg	**Lifespan** 평균수명	4~5 years
Diet 식성	Omnivore 잡식	**Habitat** 서식지	Forest 숲
Behavior 행동	Solitary living 단독생활	**Distribution** 분포지역	Thai-Malay Peninsula, Sumatra, Borneo 태국-말레이반도, 수마트라, 보르네오
Reproduction 번식	Average 1~3 평균 1~3마리	**Conservation status** 멸종위기등급	Least concern (LC) 관심대상종

LC

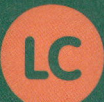

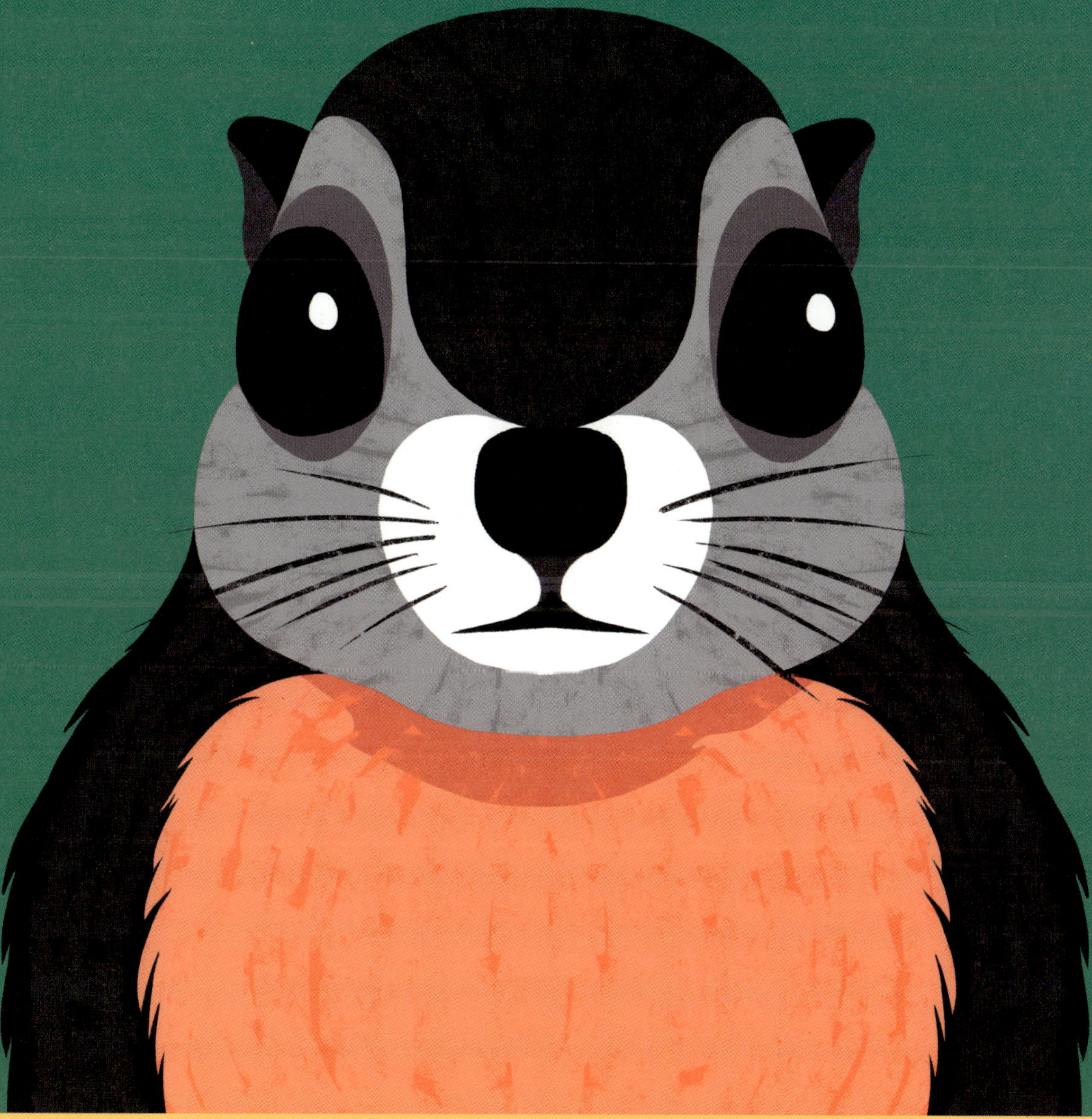

PREVOST'S SQUIRREL
LEAST CONCERN SPECIES

- EX Extinct
- EW Extinct in the Wild
- CR Critically Endangered
- EN Endangered
- VU Vulnerable
- NT Near Threatened
- LC Least Concern

WE'RE BEING MISTREATED__
PIGEON

안녕, 너희들은 나를 닭둘기라 부르지만 내 이름은 비둘기야. 우리는 하루에도 몇 번씩은 보는 사이니까 말 편하게 할게. 나도 한때는 평화와 희망의 상징이었지만 이제는 누구도 반기지 않고 마치 병균 같은 취급을 받고 있지. 최근 대한민국 환경부는 악취와 배설물 때문에 나와 내 친구들을 유해 야생 동물로 지정하기도 했어.

우리가 도시에 이렇게 많이 살게 된 이유를 알려줄게. 수십 년 전부터 대한민국은 각종 국가 행사에서 평화의 상징으로 비둘기를 날리는 연출을 많이 했어. 1988년 올림픽 당시 수천 마리의 비둘기를 하늘로 날려 보냈고 이 외에도 다양한 행사를 위한 비둘기 수요가 점차 늘어나자 수입을 하기도 하고, 국가 기관 및 공원에서 사육하기도 하고, 민간 사육을 장려하기도 했지. 나는 본능적으로 친숙한 장소로 다시 되돌아오는 귀소 본능이 강한 동물인데, 매번 행사 때마다 우리를 수천 마리씩 날려버리다 보니 이 복잡한 도시에서 그만 길을 잃고 만 거야. 우리를 잡아먹을 매, 독수리 따위가 도시에 있을리 없고 여기저기 버려진 음식물들이 많으니 우리는 기하급수적으로 가족이 늘어날 수 밖에 없었어.

Hi there, you guys call me 'chicken pigeon', but my name is actually just Pigeon. Since we see each other several times a day, let's just talk casually. I used to be a symbol of peace and hope, but now I'm treated like a germ that nobody wants around. Recently, the South Korean Ministry of Environment even designated me and my friends as 'harmful wildlife' because of our smell and excrement.

Let me explain why so many of us live in this city. For a long time, South Korea has released pigeons as a symbol of peace at various national events. During the 1988 Olympics, thousands of pigeons were released into the sky, and demand for pigeons for various events increased over time. The government even started breeding pigeons in public parks, and encouraged private breeding. I have a strong homing instinct that makes me naturally return to familiar places, but after being released in the thousands at events, I often get lost in this complicated city. There are no predators like hawks or eagles in the city to eat us, and there's plenty of discarded food everywhere, so our population has skyrocketed.

우리는 억울하다 — 비둘기

우리를 보기만 해도 마치 병균 덩어리인 마냥 혐오의 눈빛을 보내는 인간들에게 힌미디 할게. 우리도 야생 동물이다보니 병균이 전혀 없지는 않아. 하지만 단순한 날갯짓만으로는 해로운 병균을 옮기지 못해. 나는 도시에 살기 전에는 원래 깨끗한 물가에 살았고, 천적들이 내 몸에서 나는 냄새를 맡을 수 없도록 매일 목욕을 하며 털을 관리하는 깔끔한 새였어. 하지만 이 도시에서는 깨끗한 물을 접할 수 없고, 각종 사고로 인해 부리나 발가락이 잘려나가 털 관리하는 게 쉽지 않아. 게다가 이 도시의 음식들은 우리가 자연에서 먹던 것들이 아니라 인간들이 먹다 버린 음식물이라 아주 짜고 기름져서 자연스레 살도 찌게 되었어.

한때는 나도 뛰어난 탐색 능력과 귀소 본능으로 멀리 떨어진 인간들끼리 서로 메시지를 주고받을 수 있도록 도움을 주는 통신원으로 활약하며 사랑받던 시절이 있었어. 너희들의 필요에 따라 이렇게 많아져 버리고, 또 여기서 이렇게 살아야 할 수 밖에 없게 되었는데 왜 내가 미움을 받아야 하는 거지?

내가 정말 유해한 동물인 거야?

Humans often look at us with a hateful gaze, as if we're a pile of germs. I want to say to them that since we're wild animals, we do have some germs on us, but we can't transmit harmful germs just by flapping our wings. I used to live near clean water and was a clean bird that bathed and groomed myself daily to avoid being detected by predators. But in the city, clean water is scarce, and it's not easy to groom myself since I've lost some of my toes and feathers due to accidents. The food in the city isn't natural, it's often salty and oily, which makes us gain weight naturally.

There was a time when I helped humans communicate by using my excellent searching skills and homing instinct. I was even loved by humans then. But now, I've become so numerous and have to live in this city. Why do I have to be hated?

Am I really a harmful animal?

Philippine eagle,
the guardian of Philippine
필리핀의 수호자, 필리핀독수리

Size 크기	90~100cm / 6.5kg	Lifespan 평균수명	40 years
Diet 식성	Carnivore 육식	Habitat 서식지	Dipterocarp forest 딥테로카프 숲
Behavior 행동	Solitary living 단독 생활	Distribution 분포지역	Island of Luzon, Samar, Leyte 루손섬, 사마르 섬, 레이테섬
Reproduction 번식	Average 1 평균 1마리	Conservation status 멸종위기 등급	Critically endangered (CR) 위급종

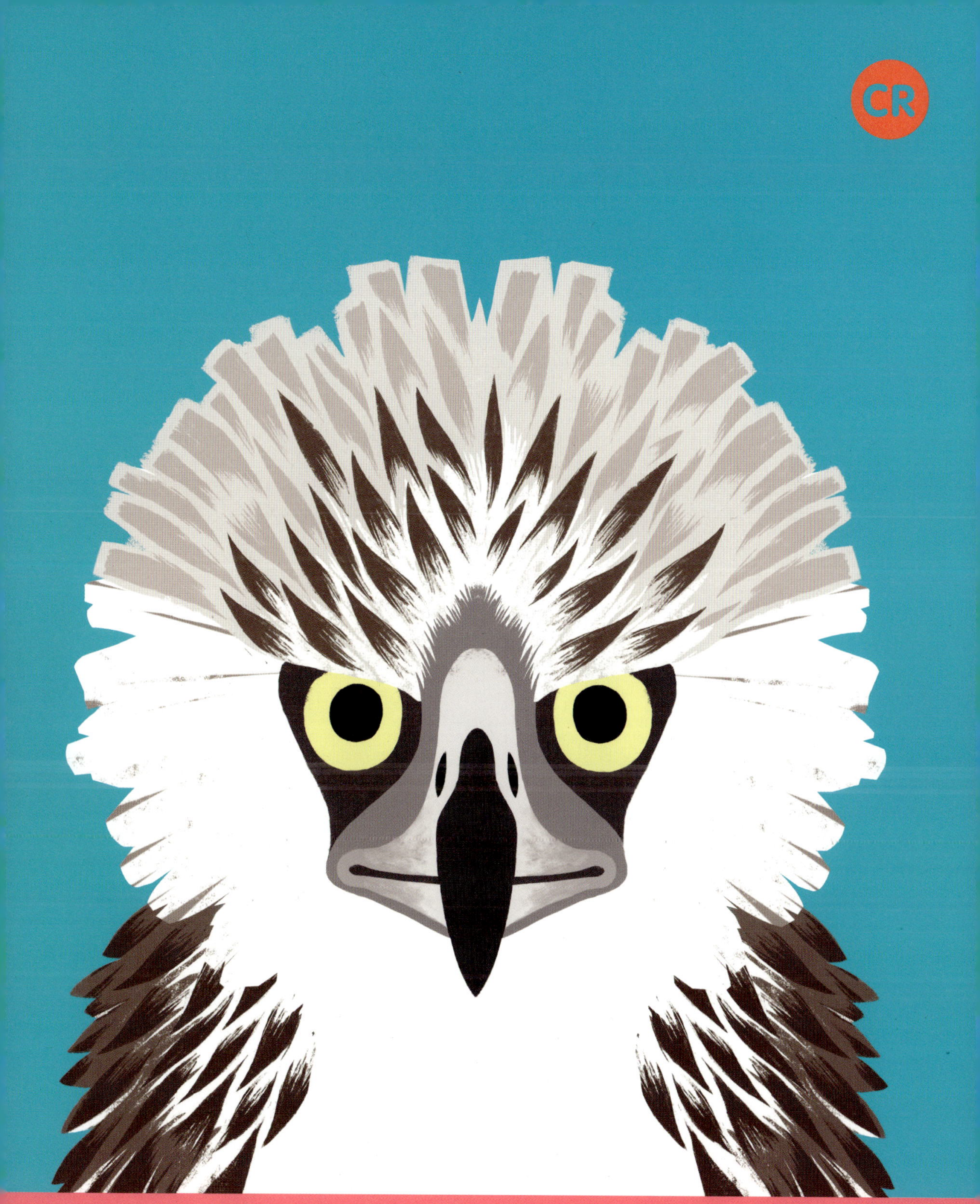

- EX Extinct
- EW Extinct in the Wild
- CR Critically Endangered
- EN Endangered
- VU Vulnerable
- NT Near Threatened
- LC Least Concern

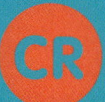

PHILIPPINE EAGLE
CRITICALLY ENDANGERED SPECIES

Common spotted cuscus,
handy tail

손 대신 꼬리, 얼룩 쿠스쿠스

Size 크기	35~45cm / 2~7kg	**Lifespan** 평균수명	11years 평균 11년
Diet 식성	Omnivore 잡식	**Habitat** 서식지	Rainforest 열대우림
Behavior 행동	Solitary living 단독생활	**Distribution** 분포지역	Cape york region of Australia, New guinea 호주 케이프요트 지역, 뉴기니
Reproduction 번식	Average 2 평균 2마리	**Conservation status** 멸종위기등급	Least concern (LC) 관심대상종

Common Spotted Cuscus has a strong tail apt to climb the trees.
얼룩쿠스쿠스는 나무타기에 좋은 강한 꼬리를 가졌습니다.

COMMON SPOTTED CUSCUS
LEAST CONCERN SPECIES

EX Extinct
EW Extinct in the Wild
CR Critically Endangered
EN Endangered
VU Vulnerable
NT Near Threatened
LC Least Concern

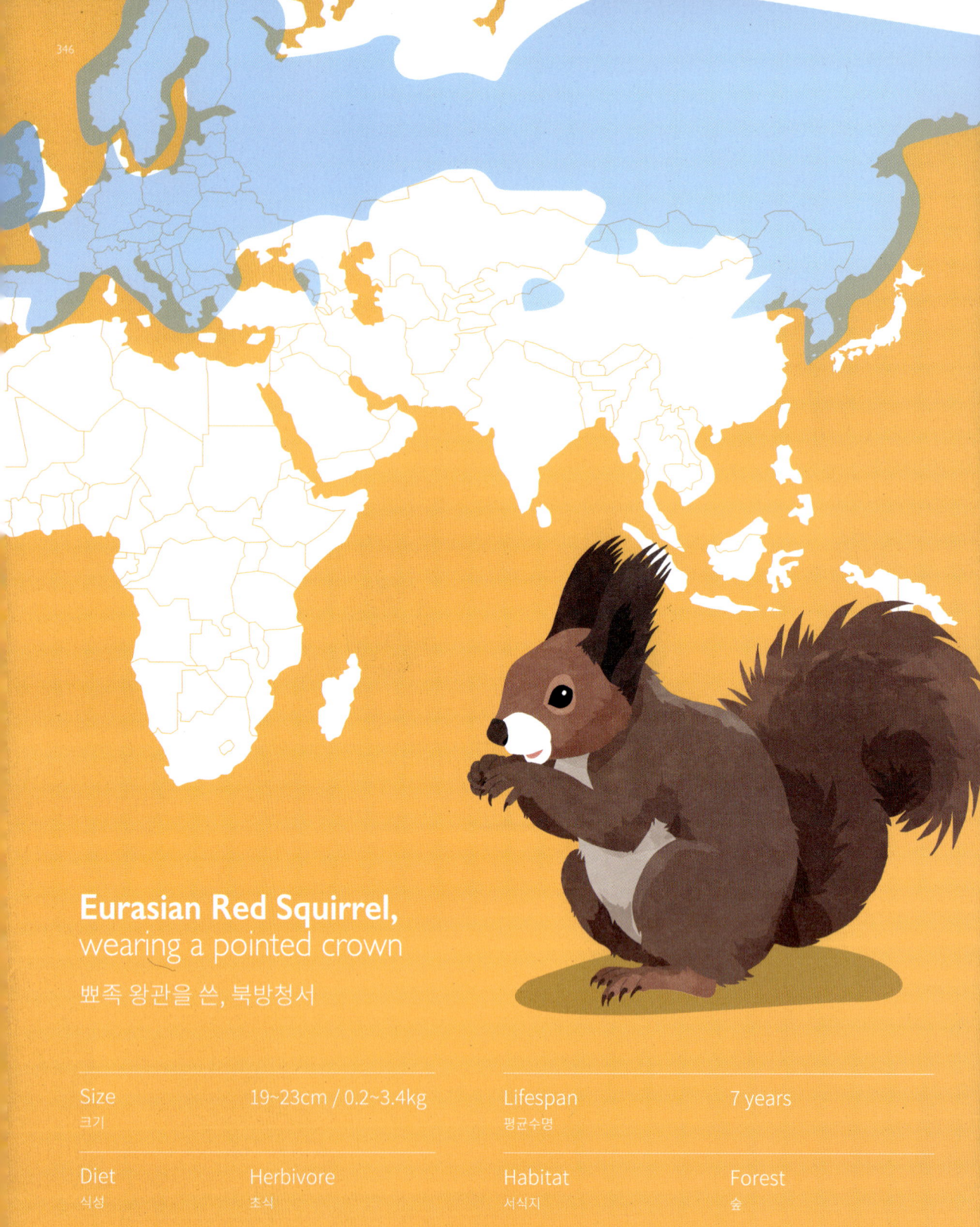

Eurasian Red Squirrel,
wearing a pointed crown
뾰족 왕관을 쓴, 북방청서

Size 크기	19~23cm / 0.2~3.4kg	**Lifespan** 평균수명	7 years
Diet 식성	Herbivore 초식	**Habitat** 서식지	Forest 숲
Behavior 행동	Solitary living 단독생활	**Distribution** 분포지역	Throughout Eurasia 유라시아 전역
Reproduction 번식	Average 3 평균 3마리	**Conservation status** 멸종위기등급	Least concern (LC) 관심대상종

Eurasian brown bear,
the violent teddybear

난폭한 테디베어, 유라시아 불곰

Size 크기	250cm / 250~300kg	Lifespan 평균수명	30 years
Diet 식성	Omnivore 잡식	Habitat 서식지	Forest 숲
Behavior 행동	Solitary living 단독 생활	Distribution 분포지역	Russia, Scandinavia, Germany 러시아, 스칸디나비아, 독일
Reproduction 번식	Average 2 평균 2마리	Conservation status 멸종위기 등급	Least concern (LC) 관심대상종

European badger,
play dead

죽은 척하기, 오소리

Size 크기	56~90cm / 6~17kg	**Lifespan** 평균수명	14 years
Diet 식성	Omnivore 잡식	**Habitat** 서식지	Savanna, forest 사바나, 숲
Behavior 행동	Group living 무리 생활	**Distribution** 분포지역	Europe, Western Asia 유럽, 서아시아 일부 지역
Reproduction 번식	Average 3 평균 3마리	**Conservation status** 멸종위기 등급	Least concern (LC) 관심대상종

An eurasian badger sometimes feigns death when in danger, such as when being chased by a natural enemy.
천적에게 쫓기는 등의 위급한 상황에서는 죽은 시늉을 하기도 합니다.

EURASIAN BADGER
LEAST CONCERN SPECIES

- **EX** Extinct
- **EW** Extinct in the Wild
- **CR** Critically Endangered
- **EN** Endangered
- **VU** Vulnerable
- **NT** Near Threatened
- **LC** Least Concern

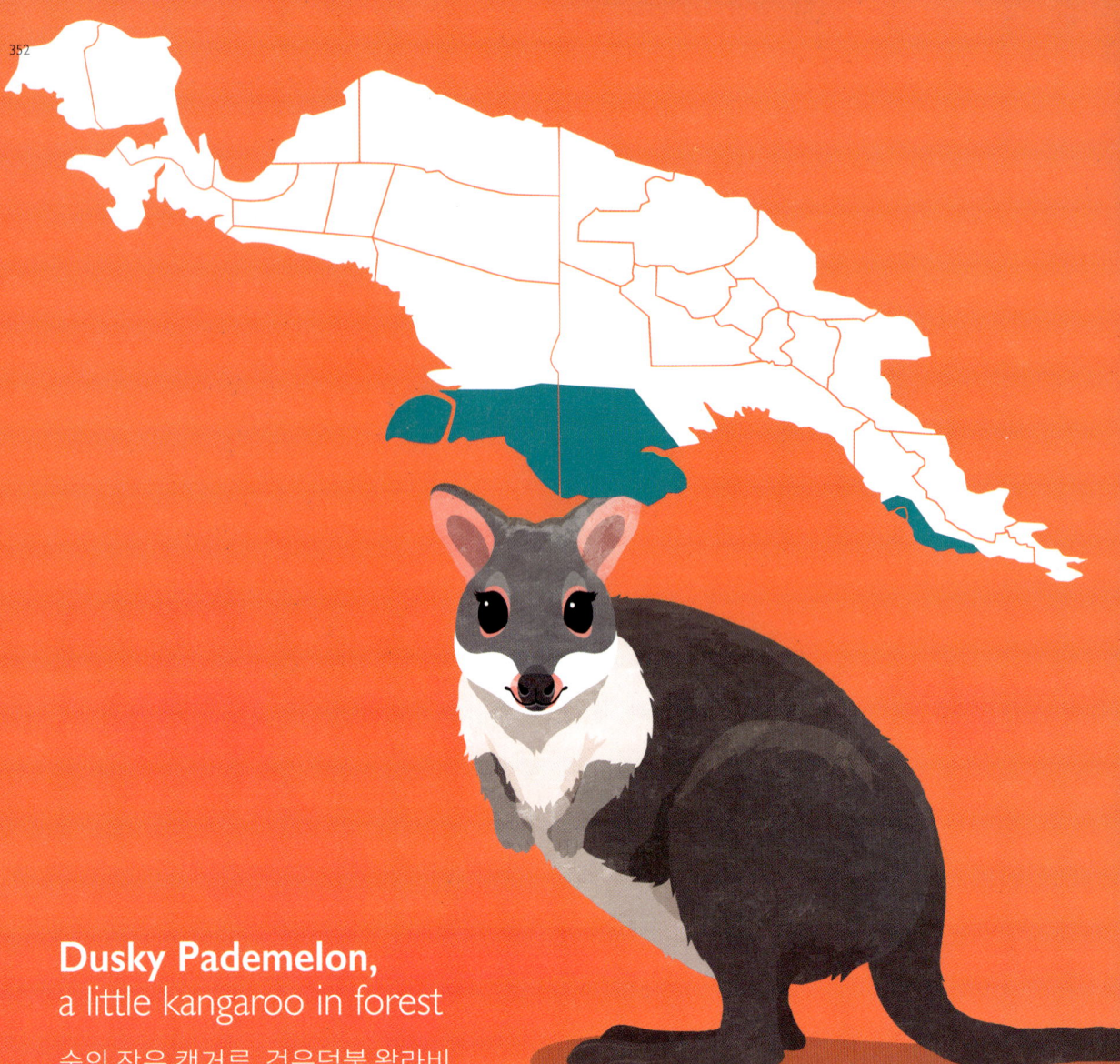

Dusky Pademelon,
a little kangaroo in forest
숲의 작은 캥거루, 검은덤불 왈라비

Size 크기	58~60cm / 3~12kg	Lifespan 평균수명	4~8 years
Diet 식성	Herbivore 초식	Habitat 서식지	Forest, Savanna 숲, 사바나
Behavior 행동	Solitary living 무리 생활	Distribution 분포지역	Indonesia, Papua New Guinea 인도네시아, 파푸아뉴기니
Reproduction 번식	Average 1 평균 1마리	Conservation status 멸종위기등급	Vulnerable (VU) 취약종

"Pademelon" is an aboriginal term from Australia that means "a small kangaroo in the forest".
"Pademelon"은 호주 원주민어로 "숲의 작은 캥거루"라는 뜻입니다.

DUSKY PADEMELON
VULNERABLE SPECIES

- EX Extinct
- EW Extinct in the Wild
- CR Critically Endangered
- EN Endangered
- **VU Vulnerable**
- NT Near Threatened
- LC Least Concern

INDIAN OCEAN

HOW TO LIVE TOGETHER

그린 인프라 :

그린 인프라는 1990년대 후반 북미에서 시작돼 세계적으로 통용되는 개념으로, 기후 변화와 자원 위기에 대응하기 위한 사회적 기반을 의미합니다. 최근에는 공원, 녹지, 하천, 텃밭, 그린벨트 등을 유기적으로 연결한 그린 네트워크를 조성해 인간과 동물이 도시에서 더불어 살아갈 수 있도록 연구하는 환경 실천 전략으로 그 개념과 정의를 확장하고 있습니다.

도시 개발에 그린 인프라가 필요한 이유는 녹지를 바탕으로 도심 환경과 생태계를 함께 유지할 때 시민들은 친환경 휴식 공간을 통해 삶의 질과 건강을 증진시키고, 동물과 식물에게 안전하고 풍요로운 서식지가 제공됨으로써 함께 살아갈 수 있기 때문입니다.

Green infrastructure :

Green infrastructure, which originated in North America in the late 1990s, is now a globally recognized concept that refers to the social foundation for addressing climate change and resource crises. Recently, the concept and definition have been expanded to include the creation of a green network that organically connects parks, green spaces, rivers, gardens, green belts, etc. to study environmental practices strategies for humans and animals to live together in cities.

The reason why green infrastructure is necessary for urban development is that when urban environments and ecosystems are maintained based on greenery, citizens can improve their quality of life and health through eco-friendly rest areas, and provide safe and abundant habitats for animals and plants to coexist.

더불어 살아가는 방법

싱가포르 정원 도시 :

싱가포르가 1965년 독립 이후 선택한 발전 전략은 청결과 녹지 조성이었습니다. 정부는 1991년부터 인구 1인당 8 ㎡의 녹지 면적을 목표로, 국가 전체의 공원을 서로 연결하여 그린 웨이를 만들겠다는 '파크 커넥터' 프로젝트를 실행합니다. 싱가포르는 국토 면적 697 ㎢로 대한민국의 부산보다 작은 국가입니다. 좁은 국토에 녹지를 효율적으로 조성하기 위한 정책으로 시작된 파크 커넥터는 크게 주거, 상업, 문화 시설을 공원과 연결시켜 시민들에게 문화 및 이동 공간을 제공하고, 그 주변으로 생태 통로를 조성해 생물 다양성을 증가시킨다는 두 가지 목적을 가지고 있습니다. 1995년 최초의 도시 공원을 시작으로 현재 300km가 넘는 파크 커넥터가 조성되었으며, 2030년까지 100km를 더 조성하는 것을 목표로 합니다.

건강한 생태계와 풍부한 먹이만 있다면 자연스럽게 동식물의 개체 수는 증가합니다. 1960년대 싱가포르에서 멸종되었던 얼룩 코뿔새는 이제 수백 마리로 늘어났으며, 대부분 자연 보호 구역뿐만이 아니라 도심을 포함한 광범위한 지역에서 자유롭게 살아갑니다. 도시공원에서 원숭이가 발견되기도 하고, 도시의 수로를 따라 수달이 보이기도 합니다. 일상에서 야생동물을 쉽게 접할 수 있는 싱가포르 시민들은 동물이 보인다고 해서 먹이를 주지 않으며 이러한 인식이 정착되어 도심 속 동식물의 생태 사이클에 인간이 방해가 되지 않도록 함께 살아가고 있습니다.

Singapore Park Connector :

Singapore's development strategy since independence in 1965 has focused on cleanliness and greenery. In 1991, the government launched the "Park Connector" project, with the goal of creating green spaces of 8 square meters per person and connecting all parks in the country. Singapore, with a land area of 697 square kilometers, is smaller than Busan, South Korea. The Park Connector project, which began as a policy to efficiently create green spaces on limited land, has two main objectives: to connect residential, commercial, and cultural facilities with parks to provide citizens with cultural and transportation spaces, and to increase biodiversity by creating ecological corridors around them. Since the first urban park was established in 1995, over 300 kilometers of Park Connectors have been created, and the goal is to create another 100 kilometers by 2030.

With a healthy ecosystem and abundant food, the number of animal populations naturally increases. The spotted wood owl, which had gone extinct in Singapore in the 1960s, now numbers in the hundreds and lives not only in nature reserves but also in extensive areas, including urban areas. Monkeys have been found in urban parks, and otters can be seen along the city's waterways. Singaporeans, who can easily encounter wildlife in their daily lives, do not feed them even when they see them, and this awareness has taken root to live in harmony with the ecological cycle of urban wildlife without disturbing them.

HOW TO LIVE TOGETHER

녹색 지붕 :

녹색 지붕이란 건물의 지붕에 식물을 심어 도시의 생태계를 조성하고 도시 환경 문제를 해결하기 위한 친환경적인 건축 기술로써 1960년대 독일에서 개발되어 여러 나라의 다양한 건축물에 적용되고 있습니다. 건물 지붕 방수막 위에 단열재, 배수층, 유지층, 흡수층 등을 쌓은 후 식물을 심어 녹지를 조성하여 인간과 동물이 도시에서 함께 살아갈 수 있는 환경을 제공합니다.

녹색 지붕에서 자란 식물이 물을 흡수, 증발하면서 발생하는 냉각 효과가 여름철 건물 온도를 낮춰주고, 겨울에는 열 보존 기능으로 실내 온도를 높여줍니다. 이러한 시스템은 열의 이동을 막아 도시의 열섬 현상을 방지하여 대기 오염을 완화시킵니다. 녹색 지붕은 도시 생태계를 보호하면서 다양한 동식물이 함께 살아갈 수 있는 환경을 제공합니다. 꽃과 나뭇잎은 벌과 나비 등 다양한 곤충을 도심으로 이끌고, 새들에게 사람의 음식물이 아닌 곤충을 잡아먹을 수 있는 환경을 제공합니다. 이처럼 녹색 지붕은 도심 속 자연 생태계를 형성해 생물 다양성을 증가시켜 지속 가능한 친환경 도시가 될 수 있게 합니다.

덴마크의 코펜하겐은 2010년부터 대형 건물에 의무적으로 녹색 지붕을 조성하도록 법으로 제정하였으며, 미국의 주요 도시들은 녹색 지붕을 여러 도시 환경과 관련된 문제점을 해결하기 위한 건축 정책중의 하나로 채택하고 있습니다.

Green Roof :

A green roof is an environmentally friendly architectural technology that involves planting vegetation on the roof of a building to create an ecosystem in the city and solve urban environmental problems. It was developed in Germany in the 1960s and has since been applied to various buildings in many countries. Layers of insulation, drainage, substrate, and vegetation are placed on top of the waterproofing layer of the roof to create a green space that provides a living environment for humans and animals in the city.

Plants grown on green roofs absorb and evaporate water, which has a cooling effect on the building during the summer and helps conserve heat in the winter. This system prevents heat island effects in the city and reduces air pollution. Green roofs protect the urban ecosystem and provide an environment where various plants and animals can coexist. Flowers and leaves attract various insects such as bees and butterflies to the city, and provide birds with an environment where they can catch insects instead of human food. In this way, green roofs create a natural ecosystem in the city, increase biodiversity, and promote sustainable, environmentally friendly cities.

Since 2010, Copenhagen, Denmark has mandated the creation of green roofs on large buildings through legislation, and major cities in the United States have adopted green roofs as one of their architectural policies to address various urban environmental issues.

더불어 살아가는 방법

대한민국 녹색 다리, 생태통로 :

생태통로란 도로, 댐 등으로 인해 단절된 생태계를 연결하고 야생 동물의 이동을 위한 인공구조물입니다. 야생 동물이 서식지를 이동할 때 도로나 철도를 거치지 않고 건널 수 있게 하며, 일반적으로 육교형 또는 터널형 다리로 건설되어 토지, 수로, 산림 등 다른 생태계를 서로 이어주는 역할을 합니다. 생태통로는 로드킬 예방 뿐만 아니라 지역 생태계와 생물다양성을 유지시키는 역할을 합니다. 도로나 철도가 동물들의 서식지를 고립시키는 경우 지역 생태계는 큰 위협을 받을 수 있습니다. 야생 동물이 서식지를 이동하며 배설물을 통해 씨앗을 퍼트려 식물의 번식이 이어지는데, 도로 건설로 인해 동물이 고립되는 경우 식물의 분산과 번식이 중단되고 결국 야생 동물의 멸종과 숲 생태계 파괴로 이어질 수 있습니다. 따라서 최대한 많은 동물이 이동할 수 있도록 지역별로 서식중인 동물의 다양성을 고려해 규모와 모양을 각각 다르게 설계합니다.

대한민국에는 현재 500여 개의 생태통로가 설치되어 있으며 추풍령 휴게소 생태통로의 현황을 조사한 결과 고니, 수달, 담비, 산양, 하늘다람쥐, 반달가슴곰 등 멸종 위기동물들도 활발히 이용하는 것이 확인되었습니다. 또한 생태통로 이용률을 분석한 결과 2014년부터 2018년까지 5년간 야생동물의 평균 이용률은 약 2.5배 증가하였습니다.

South Korea, Eco Bridge :

An ecological corridor is an artificial structure such as a road or dam that connects fragmented ecosystems and facilitates the movement of wildlife. Typically, eco-corridors are constructed in the form of pedestrian bridges or tunnels to allow animals to cross over without having to go through roads or railways, and they serve to link different ecosystems such as land, water, and forests. Eco-corridors play a crucial role not only in preventing roadkill accidents but also in maintaining local ecosystems and biodiversity. When roads or railways isolate animal habitats, the local ecosystem is put at great risk. Wild animals move around and disperse seeds through their excrement, which is vital to the reproduction of plants. When animals become isolated due to road construction, plant distribution and reproduction can be disrupted, eventually leading to the extinction of wild animals and destruction of forest ecosystems. Therefore, to enable as many animals as possible to move around, eco-corridors are designed with different scales and shapes, taking into account the diversity of animal species inhabiting the area.

In South Korea, there are currently over 500 eco-corridors installed, and a survey of the Chupungryeong rest area eco-corridor found that even endangered species such as goral, otters, pheasants, mountain goats, sky squirrels, and Asiatic black bears are actively utilizing the corridor. An analysis of the eco-corridor usage rate revealed that the average usage rate of wild animals increased by about 2.5 times from 2014 to 2018.

REFERENCE
참고자료

10p -BirdLife International (2016). "Aegolius acadicus". IUCN Red List of Threatened Species. 2016: e.T22689366A93228694. doi:10.2305/IUCN.UK.2016-3.RLTS.T22689366A93228694.en. Retrieved 11 November 2021.
-Beolens, Bo; Watkins, Michael; Grayson, Michael (2009). The Eponym Dictionary of Mammals. Baltimore: Johns Hopkins University Press. p. 220. ISBN 9780801895333.

12p - Ravetta, A.L.; Calouro, A.M.; Wallace, R.B.; Mollinedo, J.M.; Röhe, F.; Bicca-Marques, J.C.; Heymann, E.W.; Mittermeier, R.A. 2021. "Saguinus imperator". IUCN Red List of Threatened Species. 2021: e.T39948A192551472. doi:10.2305/IUCN.UK.2021-1.RLTS.T39948A192551472.en. Retrieved 12 November 2021.
- Hershkovitz Philip. 1979. "Races of the emperor tamarin, Saguinus imperator Goeldi Callitrichidae, Primates"

14p - Gardner, A. L. 2005. "Genus Choloepus". In Wilson, D. E.; Reeder, D. M. eds.. Mammal Species of the World: A Taxonomic and Geographic Reference 3rd ed.. Johns Hopkins University Press. pp. 101–102. ISBN 978-0-8018-8221-0. OCLC 62265494.
- Delsuc, F.; Kuch, M.; Gibb, G. C.; Karpinski, E.; Hackenberger, D.; Szpak, P.; Martínez, J. G.; Mead, J. I.; McDonald, H. G.; MacPhee, R.D.E.; Billet, G.; Hautier, L.; Poinar, H. N. 2019. "Ancient Mitogenomes Reveal the Evolutionary History and Biogeography of Sloths". Current Biology. 29 12: 2031–2042.e6. doi:10.1016/j.cub.2019.05.043. PMID 31178321.

16p - Groves, C. P. 2005. Wilson, D. E.; Reeder, D. M. eds.. Mammal Species of the World: A Taxonomic and Geographic Reference 3rd ed.. Baltimore: Johns Hopkins University Press. p. 131. ISBN 0-801-88221-4. OCLC 62265494.
- Rylands AB, Mittermeier RA 2009. "The Diversity of the New World Primates Platyrrhini". In Garber PA, Estrada A, Bicca-Marques JC, Heymann EW, Strier KB eds.. South American Primates: Comparative Perspectives in the Study of Behavior, Ecology, and Conservation. Springer. pp. 23–54. ISBN 978-0-387-78704-6.

18p - Wozencraft, W. C. 2005. "Order Carnivora". In Wilson, D. E.; Reeder, D. M. eds.. Mammal Species of the World: A Taxonomic and Geographic Reference 3rd ed.. Johns Hopkins University Press. pp. 627–628. ISBN 978-0-8018-8221-0. OCLC 62265494.
- Cuarón, A.D.; de Grammont, P.C.; McFadden, K. 2016. "Procyon pygmaeus". IUCN Red List of Threatened Species. 2016: e.T18267A45201913. doi:10.2305/IUCN.UK.2016-1.RLTS.T18267A45201913.en. Retrieved 12 November 2021.

20p - British Museum. "King penguin: The Forsters, King and Emperor". Explore/Highlights. Trustees of the British Museum. Archived from the original on 5 August 2008. Retrieved 28 July 2013.
- BirdLife International 2020. "Aptenodytes forsteri". IUCN Red List of Threatened Species. 2020: e.T22697752A157658053. doi:10.2305/IUCN.UK.2020-3.RLTS.T22697752A157658053.en. Retrieved 19 November 2021.

22p - Belant, J.; Biggins, D.; Garelle, D.; Griebel, R.G. & Hughes, J.P. 2015. "Mustela nigripes". IUCN Red List of Threatened Species. 2015: e.T14020A45200314. doi:10.2305/IUCN.UK.2015-4.RLTS.T14020A45200314.en. Retrieved February 8, 2022.
- Heptner, V. G. Vladimir Georgievich; Nasimovich, A. A; Bannikov, Andrei Grigorovich; Hoffmann, Robert S. 2001. Mammals of the Soviet Union Volume: v. 2, pt. 1b. Washington, D.C. : Smithsonian Institution Libraries and National Science Foundation.

26p - Groves, C. P. (2005). Wilson, D. E.; Reeder, D. M. (eds.). Mammal Species of the World: A Taxonomic and Geographic Reference (3rd ed.). Baltimore: Johns Hopkins University Press. p. 135. ISBN 0-801-88221-4. OCLC 62265494.
- V., Link, A., Guzman-Caro, D., Defler, T.R., Palacios, E., Stevenson, P.R. & Mittermeier, R.A. 2021. Saguinus oedipus (amended version of 2020 assessment). The IUCN Red List of Threatened Species 2021: e.T19823A192551067. https://dx.doi.org/10.2305/IUCN.UK.2021-1.RLTS.T19823A192551067.en. Downloaded on 06 April 2021.

28p - Wozencraft, W. C. 2005. "Order Carnivora". In Wilson, D. E.; Reeder, D. M. eds.. Mammal Species of the World: A Taxonomic and Geographic Reference 3rd ed.. Johns Hopkins University Press. pp. 539–540. ISBN 978-0-8018-8221-0. OCLC 62265494.

- de Oliveira, T.; Paviolo, A.; Schipper, J.; Bianchi, R.; Payan, E. & Carvajal, S.V. 2015. "Leopardus wiedii". IUCN Red List of Threatened Species. 2015: e.T11511A50654216. doi:10.2305/IUCN.UK.2015-4.RLTS.T11511A50654216.en. Retrieved 16 January 2022.

30p - BirdLife International 2020. "Eudyptes moseleyi". IUCN Red List of Threatened Species. 2020: e.T22734408A184698049. doi:10.2305/IUCN.UK.2020-3.RLTS.T22734408A184698049.en. Retrieved 11 November 2021.
- Jouventin, P; Cuthbert, R. J.; Ottvall, R. 2006. "Genetic isolation and divergence in sexual traits: Evidence for the northern rockhopper penguin Eudyptes moseleyi being a sibling species". Molecular Ecology. 15 11: 3413–3423. doi:10.1111/j.1365-294X.2006.03028.x. PMID 16968279. S2CID 12724253.

32p - Gardner, A. L. 2005. "Order Pilosa". In Wilson, D. E.; Reeder, D. M. eds.. Mammal Species of the World: A Taxonomic and Geographic Reference 3rd ed.. Johns Hopkins University Press. pp. 100–101. ISBN 978-0-8018-8221-0. OCLC 62265494.
- Delsuc, F.; Kuch, M.; Gibb, G.C.; Karpinski, E.; Hackenberger, D.; Szpak, P.; et al. 2019. "Ancient Mitogenomes reveal the evolutionary history and biogeography of sloths". Current Biology. 29 12: 2031–2042.e6. doi:10.1016/j.cub.2019.05.043. PMID 31178321.

34p - Roach, N. 2017. "Marmota vancouverensis". IUCN Red List of Threatened Species. 2017: e.T12828A22259184. doi:10.2305/IUCN.UK.2017-2.RLTS.T12828A22259184.en. Retrieved 12 November 2021.
-Canada, Environment and Climate Change. "Vancouver Island Marmot Marmota vancouverensis: COSEWIC assessment and status report 2019". www.canada.ca. Retrieved 2022-09-22.

36p - Kovacs, K.M. 2015. "Pagophilus groenlandicus". IUCN Red List of Threatened Species. 2015: e.T41671A45231087. doi:10.2305/IUCN.UK.2015-4.RLTS.T41671A45231087.en. Retrieved 4 April 2022.
- Encyclopedia of marine mammals. Perrin, William F., Würsig, Bernd G., Thewissen, J. G. M. 2nd ed.. Amsterdam: Elsevier/Academic Press. 2009. ISBN 9780123735539. OCLC 316226747.

38p - Groves, C. P. 2005. Wilson, D. E.; Reeder, D. M. eds.. Mammal Species of the World: A Taxonomic and Geographic Reference 3rd ed.. Baltimore: Johns Hopkins University Press. p. 133. ISBN 0-801-88221-4. OCLC 62265494.
- Rylands AB, Mittermeier RA 2009. "The Diversity of the New World Primates Platyrrhini". In Garber PA, Estrada A, Bicca-Marques JC, Heymann EW, Strier KB eds.. South American Primates: Comparative Perspectives in the Study of Behavior, Ecology, and Conservation. Springer. pp. 23–54. ISBN 978-0-387-78704-6.

42p - Groves, C. P. 2005. Wilson, D. E.; Reeder, D. M. eds.. Mammal Species of the World: A Taxonomic and Geographic Reference 3rd ed.. Baltimore: Johns Hopkins University Press. p. 133. ISBN 0-801-88221-4. OCLC 62265494.
- Rylands AB, Mittermeier RA 2009. "The Diversity of the New World Primates Platyrrhini". In Garber PA, Estrada A, Bicca-Marques JC, Heymann EW, Strier KB eds.. South American Primates: Comparative Perspectives in the Study of Behavior, Ecology, and Conservation. Springer. pp. 23–54. ISBN 978-0-387-78704-6.

44p - Kurtén, Björn (1968). Pleistocene Mammals of Europe. Transaction Publishers. pp. 170–177. ISBN 978-1-4128-4514-4. Retrieved 6 August 2013.
- Harding, Lee E. (26 August 2022). "Available names for Rangifer (Mammalia, Artiodactyla, Cervidae) species and subspecies". ZooKeys (1119): 117–151. doi:10.3897/zookeys.1119.80233. ISSN 1313-2970.

46p - Helgen, K.; Reid, F. 2016. "Taxidea taxus". IUCN Red List of Threatened Species. 2016: e.T41663A45215410. doi:10.2305/IUCN.UK.2016-1.RLTS.T41663A45215410.en. Retrieved November 19, 2021.
- Wozencraft, W. C. 2005. "Order Carnivora". In Wilson, D. E.; Reeder, D. M. eds.. Mammal Species of the World: A Taxonomic and Geographic Reference 3rd ed.. Johns Hopkins University Press. p. 619. ISBN 978-0-8018-8221-0. OCLC 62265494.

48p - Stevenson, P.R., Defler, T.R., de la Torre, S., Moscoso, P., Palacios, E., Ravetta, A.L., Vermeer, J., Link, A., Urbani, B., Cornejo, F.M., Guzmán-Caro, D.C., Shanee, S., Mourthé, Í., Muniz, C.C., Wallace, R.B. & Rylands, A.B. (2021). "Lagothrix lagotricha". The IUCN Red List of Threatened Species. IUCN. 2021: e.T160881218A192309103. doi:10.2305/IUCN.UK.2021-1.RLTS.T160881218A192309103.en.
- "Oldstyle id: c34625c16245785ce9b441b53e92475a". Species 2000 & ITIS Catalogue of Life. Species 2000: Naturalis, Leiden, the Netherlands.

50p - Hoffmann, M. & Sillero-Zubiri, C. 2021 [amended version of 2016 assessment]. "Vulpes vulpes". IUCN Red List of Threatened Species. 2021: e.T23062A193903628.

doi:10.2305/IUCN.UK.2021-1.RLTS.T23062A193903628.en. Retrieved 17 February 2022.
- Linnaeus, C. 1758. "Canis Vulpes". Caroli Linnæi Systema naturæ per regna tria naturæ, secundum classes, ordines, genera, species, cum characteribus, differentiis, synonymis, locis in Latin. Vol. Tomus I decima, reformata ed.. Holmiae: Laurentius Salvius. p. 40.

52p - Cuarón, A.D.; Helgen, K.; Reid, F.; Pino, J.; González-Maya, J.F. (2016). "Nasua narica". IUCN Red List of Threatened Species. 2016: e.T41683A45216060.
doi:10.2305/IUCN.UK.2016-1.RLTS.T41683A45216060.en. Retrieved 19 November 2021.
- Wilson, D.E.; Reeder, D.M., eds. (2005). "Species Nasua narica". Mammal Species of the World: A Taxonomic and Geographic Reference (3rd ed.). Johns Hopkins University Press. ISBN 978-0-8018-8221-0. OCLC 62265494.

54p - Rodriguez, B. and Pineda, W. (2015). "Ectophylla alba". IUCN Red List of Threatened Species. 2015: e.T7030A22027138.
doi:10.2305/IUCN.UK.2015-4.RLTS.T7030A22027138.en. Retrieved 19 November 2021.
- Gillam, Erin H; Chaverri, Gloriana; Montero, Karina; Sagot, Maria (2013). "Social Calls Produced within and near the Roost in Two Species of Tent-Making Bats, Dermanura watsoni and Ectophylla alba". PLOS ONE. 8 (4): e61731. Bibcode:2013PLoSO...861731G. doi:10.1371/journal.pone.0061731. PMC 3634860. PMID 23637893.

58p - Reid, F. (2016). "Hydrochoerus hydrochaeris". IUCN Red List of Threatened Species. 2016: e.T10300A22190005.
doi:10.2305/IUCN.UK.2016-2.RLTS.T10300A22190005.en. Retrieved 19 November 2021.
- Capybara (Hydrochoerus hydrochaeris) Archived 2012-01-03 at the Wayback Machine. ARKive.org

60p - Angerbjörn, A. & Tannerfeldt, M. 2014. "Vulpes lagopus". IUCN Red List of Threatened Species. 2014: e.T899A57549321.
doi:10.2305/IUCN.UK.2014-2.RLTS.T899A57549321.en. Retrieved 19 November 2021.
- Linnæus, C. 1758. "Vulpes lagopus". Systema naturæ per regna tria naturæ, secundum classes, ordines, genera, species, cum characteribus, differentiis, synonymis, locis. Tomus I in Latin 10th ed.. Holmiæ Stockholm: Laurentius Salvius. p. 40. Archived from the original on 8 November 2012. Retrieved 23 November 2012.

62p - Valença-Montenegro, M.M.; Bezerra, B.M.; Martins, A.B.; Jerusalinsky, L.; Fialho, M.S.; Lynch Alfaro, J.W. (2021). "Sapajus flavius". IUCN Red List of Threatened Species. 2021: e.T136253A192592928.
doi:10.2305/IUCN.UK.2021-1.RLTS.T136253A192592928.en. Retrieved 19 November 2021.
- Marcgrave, G. (1648). "Liber sextus: De quadrupedibus, et sepentibus". Historia Naturalis Brasiliae. Lugdunum Batavorum: Franciscus Hackius. pp. 226–227., cited by de Oliveira & Langguth 2006.

64p - Lizcano, D.J.; Amanzo, J.; Castellanos, A.; Tapia, A.; Lopez-Malaga, C.M. (2016). "Tapirus pinchaque". IUCN Red List of Threatened Species. 2016: e.T21473A45173922.
doi:10.2305/IUCN.UK.2016-1.RLTS.T21473A45173922.en. Retrieved 19 November 2021.
- Roulin, F. (1829). "Mémoire pour servir à l'histoire du Tapir; et Description d'une espèce nouvelle appartenant aux hautes régions de la Cordillère des Andes". Annales des sciences naturelles. 18: 26–56.

66p - Baldi, R.B.; Acebes, P.; Cuéllar, E.; Funes, M.; Hoces, D.; Puig, S.; Franklin, W.L. (2016). "Lama guanicoe". IUCN Red List of Threatened Species. 2016: e.T11186A18540211.
doi:10.2305/IUCN.UK.2016-1.RLTS.T11186A18540211.en. Retrieved 19 November 2021.
- "Guanaco – LAMA GUANICOE". America Zoo. Lesley Fountain. Archived from the original on 28 April 2009.

68p - "jaguarundi". Lexico UK English Dictionary. Oxford University Press. Archived from the original on October 17, 2021.
- Caso, A.; de Oliveira, T. & Carvajal, S.V. (2015). "Herpailurus yagouaroundi". IUCN Red List of Threatened Species. 2015: e.T9948A50653167.
doi:10.2305/IUCN.UK.2015-2.RLTS.T9948A50653167.en. Retrieved 15 January 2022.

70p - Hückstädt, L. 2015. "Leptonychotes weddellii". IUCN Red List of Threatened Species. 2015: e.T11696A45226713.
doi:10.2305/IUCN.UK.2015-4.RLTS.T11696A45226713.en. Retrieved 19 November 2021.
- Wozencraft, W. C. 2005. "Order Carnivora". In Wilson, D. E.; Reeder, D. M. eds.. Mammal Species of the World: A Taxonomic and Geographic Reference 3rd ed.. Johns Hopkins University Press. ISBN 978-0-8018-8221-0. OCLC 62265494.

74p - Link, A.; Urbani, B.; Mittermeier, R.A. (2021). "Aotus griseimembra". IUCN Red List of Threatened Species. 2021: e.T1807A190452803.
doi:10.2305/IUCN.UK.2021-1.RLTS.T1807A190452803.en. Retrieved 20 November 2021.
- Groves, C. P. (2005). Wilson, D. E.; Reeder, D. M. (eds.). Mammal Species of the World: A Taxonomic and Geographic Reference (3rd ed.). Baltimore: Johns Hopkins University Press. p. 140. ISBN 0-801-88221-4. OCLC 62265494.

76p - Groves, C. P. 2005. Wilson, D. E.; Reeder, D. M. eds.. Mammal Species of the World: A Taxonomic and Geographic Reference 3rd ed.. Baltimore: Johns Hopkins University Press. pp. 146–148. ISBN 0-801-88221-4. OCLC 62265494.
- Aquino, R.; de Queiroz, H.L.; Paim, F.P.; Boubli, J.P.; Mittermeier, R.A.; Ravetta, A.L.; Shanee, S.; Urbani, B.; de Azevedo, R.B.; Calouro, A.M.; Cornejo, F.M. 2021 [amended version of 2020 assessment]. "Cacajao calvus". IUCN Red List of Threatened Species. 2021: e.T3416A191694447.
doi:10.2305/IUCN.UK.2021-1.RLTS.T3416A191694447.en. Retrieved 12 March 2022.

78p - Gompper, M.; Jachowski, D. 2016. "Spilogale putorius". IUCN Red List of Threatened Species. 2016: e.T41636A45211474.
doi:10.2305/IUCN.UK.2016-1.RLTS.T41636A45211474.en. Retrieved 13 November 2021.
- "NatureServe Explorer 2.0". explorer.natureserve.org. Retrieved 4 April 2022.

80p - Velez-Liendo, X.; García-Rangel, S. 2018 [errata version of 2017 assessment]. "Tremarctos ornatus". IUCN Red List of Threatened Species. 2017: e.T22066A123792952.
doi:10.2305/IUCN.UK.2017-3.RLTS.T22066A45034047.en. Retrieved 14 February 2020.
- Sichra, Inge 2003. La vitalidad del quechua: lengua y sociedad en dos provincias de Cochabamba in Spanish. Plural editores. p. 121. ISBN 9789990575149.

82p - Gardner, A. L. 2005. "Species Cyclopes didactylus". In Wilson, D. E.; Reeder, D. M. eds.. Mammal Species of the World: A Taxonomic and Geographic Reference 3rd ed.. Johns Hopkins University Press. p. 102. ISBN 978-0-8018-8221-0. OCLC 62265494.
- Miranda, F.; Meritt, D.A.; Tirira, D.G.; Arteaga, M. 2014. "Cyclopes didactylus". IUCN Red List of Threatened Species. 2014: e.T6019A47440020.
doi:10.2305/IUCN.UK.2014-1.RLTS.T6019A47440020.en. Retrieved 19 November 2021.

84p - Wozencraft, W. C. 2005. "Species Leopardus jacobitus". In Wilson, D. E.; Reeder, D. M. eds.. Mammal Species of the World: A Taxonomic and Geographic Reference 3rd ed. Johns Hopkins University Press. pp. 532–628. ISBN 978-0-8018-8221-0. OCLC 62265494.
- Villalba, L.; Lucherini, M.; Walker, S.; Lagos, N.; Cossios, D.; Bennett, M. & Huaranca, J. 2016. "Leopardus jacobita". IUCN Red List of Threatened Species. 2016: e.T15452A50657407.
doi:10.2305/IUCN.UK.2016-1.RLTS.T15452A50657407.en. Retrieved 15 January 2022.

86p - Reid, F.; Schipper, J. & Timm, R. (2016). "Bassariscus astutus". IUCN Red List of Threatened Species. 2016: e.T41680A45215881.
doi:10.2305/IUCN.UK.2016-1.RLTS.T41680A45215881.en. Retrieved November 19, 2021.
- State mammal. Arizona State Library, Archives, & Public Records (Report). State of Arizona. Retrieved May 24, 2019.

90p - Wozencraft, W. C. 2005. "Order Carnivora". In Wilson, D. E.; Reeder, D. M. eds.. Mammal Species of the World: A Taxonomic and Geographic Reference 3rd ed.. Johns Hopkins University Press. p. 583. ISBN 978-0-8018-8221-0. OCLC 62265494.
- Coonan, T.; Ralls, K.; Hudgens, B.; Cypher, B.; Boser, C. 2013. "Urocyon littoralis". IUCN Red List of Threatened Species. 2013: e.T22781A13985603.
doi:10.2305/IUCN.UK.2013-2.RLTS.T22781A13985603.en. Retrieved 9 March 2022.

92p - Festa-Bianchet, M. 2020. "Ovis canadensis". IUCN Red List of Threatened Species. 2020: e.T15735A22146699.
doi:10.2305/IUCN.UK.2020-2.RLTS.T15735A22146699.en. Retrieved 12 November 2021.
- "Ovis canadensis". The IUCN Red List of Threatened Species. IUCN International Union for Conservation of Nature. 2008. Version 2016-2. Retrieved 2016-11-20.

94p - Hoffman, R.S.; Smith, A.T. 2005. "Order Lagomorpha". In Wilson, D.E.; Reeder, D.M eds.. Mammal Species of the World: A Taxonomic and Geographic Reference 3rd ed.. Johns Hopkins University Press. p. 199. ISBN 978-0-8018-8221-0. OCLC 62265494.
- Cervantes, F.A.; Lorenzo, C.; Farías, V. & Vargas, J. 2008. "Lepus flavigularis". The IUCN Red List of Threatened Species. IUCN. 2008: e.T11790A3306162.

doi:10.2305/IUCN.UK.2008.RLTS.T11790A3306162.en. Retrieved 24 December 2017.
96p - Trillmich, F. (2015). "Arctocephalus galapagoensis". IUCN Red List of Threatened Species. 2015: e.T2057A45223722. doi:10.2305/IUCN.UK.2015-2.RLTS.T2057A45223722.en. Retrieved 19 November 2021.
- "Galapagos Fur Seals ~ MarineBio Conservation Society". 18 May 2017. Retrieved 2021-11-28.
98p - Gardner, A. L. (2005). "Order Pilosa". In Wilson, D. E.; Reeder, D. M. (eds.). Mammal Species of the World: A Taxonomic and Geographic Reference (3rd ed.). Johns Hopkins University Press. p. 103. ISBN 978-0-8018-8221-0. OCLC 62265494.
- Miranda, F.; Fallabrino, A.; Arteaga, M.; Tirira, D.G.; Meritt, D.A.; Superina, M. (2014). "Tamandua tetradactyla". IUCN Red List of Threatened Species. 2014: e.T21350A47442916. doi:10.2305/IUCN.UK.2014-1.RLTS.T21350A47442916.en. Retrieved 12 November 2021.
100p - Wozencraft, W. C. (2005). "Order Carnivora". In Wilson, D. E.; Reeder, D. M. (eds.). Mammal Species of the World: A Taxonomic and Geographic Reference (3rd ed.). Johns Hopkins University Press. pp. 532–628. ISBN 978-0-8018-8221-0. OCLC 62265494.
- Paula, R.C.; DeMatteo, K. (2016) [errata version of 2015 assessment]. "Chrysocyon brachyurus". IUCN Red List of Threatened Species. 2015: e.T4819A88135664. doi:10.2305/IUCN.UK.2015-4.RLTS.T4819A82316878.en. Retrieved 18 February 2022.
102p - BirdLife International 2018. "Ara glaucogularis". IUCN Red List of Threatened Species. 2018: e.T22685542A130868462. doi:10.2305/IUCN.UK.2018-2.RLTS.T22685542A130868462.en. Retrieved 12 November 2021.
- "Ley N° 584 – Declara Patrimonio Natural del Estado Plurinacional de Bolivia, a la Paraba Barba Azul Ara glaucogularis". www.ecolex.org in Spanish. Archived from the original on 2017-12-22. Retrieved 2017-12-20.
106p - de la Torre, S.; Shanee, S.; Palacios, E.; Calouro, A.M.; Messias, M.R.; Valença-Montenegro, M.M. (2021). "Cebuella pygmaea". IUCN Red List of Threatened Species. 2021: e.T136926A200203263. doi:10.2305/IUCN.UK.2021-2.RLTS.T136926A200203263.en. Retrieved 18 November 2021.
- Genoud, Michel; Martin, Robert D.; Glaser, Dieter (1997). <229::aid-ajp5>3.0.co;2-z "Rate of metabolism in the smallest simian primate, the pygmy marmoset (Cebuella pygmaea)". American Journal of Primatology. 41 (3): 229–245. doi:10.1002/(sici)1098-2345(1997)41:3<229::aid-ajp5>3.0.co;2-z. ISSN 0275-2565. PMID 9057967. S2CID 20927342.
108p - Defler, Thomas R.; Bueno, Marta; Garcia, Javier (2010). "Callicebus caquetensis: A New and Critically Endangered Titi Monkey from Southern Caquetá, Colombia". Primate Conservation. 25: 1–9. doi:10.1896/052.025.0101. S2CID 83583912. 136211. Archived from the original on 2011-07-26. Retrieved 2010-07-18.
- Defler, T.R.; García-Gutiérrez, J.; Guzmán-Caro, D.C.; Palacios, E.; Stevenson, P.R. (2021). "Plecturocebus caquetensis". IUCN Red List of Threatened Species. 2021: e.T14699281A192453101. doi:10.2305/IUCN.UK.2021-1.RLTS.T14699281A192453101.en. Retrieved 12 November 2021.
110p - Helgen, K.; Reid, F. (2016). "Martes americana". IUCN Red List of Threatened Species. 2016: e.T41648A45212861. doi:10.2305/IUCN.UK.2016-1.RLTS.T41648A45212861.en. Retrieved 19 February 2022.
- Wilson, D.E.; Reeder, D.M., eds. (2005). "Martes americana". Mammal Species of the World: A Taxonomic and Geographic Reference (3rd ed.). Johns Hopkins University Press. ISBN 978-0-8018-8221-0. OCLC 62265494.
112p - BirdLife International 2019. "Tyto alba". IUCN Red List of Threatened Species. 2019: e.T22688504A155542941. doi:10.2305/IUCN.UK.2019-3.RLTS.T22688504A155542941.en. Retrieved 13 November 2021.
- Peters, James Lee 1964. Check-list of Birds of the World. Volume IV. Harvard University Press. p. 77–82.
114p - Doroff, A.; Burdin, A. 2015. "Enhydra lutris". IUCN Red List of Threatened Species. 2015: e.T7750A21939518. doi:10.2305/IUCN.UK.2015-2.RLTS.T7750A21939518.en. Retrieved 11 November 2021.
- Womble, Jamie 29 July 2016. "A Keystone Species, the Sea Otter, Colonizes Glacier Bay". National Park Service. Retrieved 23 November 2021.
116p - Groves, C. P. 2005. Wilson, D. E.; Reeder, D. M. eds.. Mammal Species of the World: A Taxonomic and Geographic Reference 3rd ed.. Baltimore: Johns Hopkins University Press. pp. 138–139. ISBN 0-801-88221-4. OCLC 62265494.
- Simpson, George Gaylord 1941. "Vernacular Names of South American Mammals". Journal of Mammalogy. 22 1: 1–17. doi:10.2307/1374677. JSTOR 1374677.
120p - Abernethy, K.; Maisels, F. 2019. "Mandrillus sphinx". IUCN Red List of Threatened Species. 2019: e.T12754A17952325. doi:10.2305/IUCN.UK.2019-3.RLTS.T12754A17952325.en. Retrieved 19 November 2021.
- Linné, C. v. 1758. "Simia sphinx". Systema naturæ. Regnum animale. Vol. 1 10th ed.. Sumptibus Guilielmi Engelmann. p. 25.
122p - BirdLife International 2020. "Terathopius ecaudatus". IUCN Red List of Threatened Species. 2020: e.T22695289A174413323. doi:10.2305/IUCN.UK.2020-3.RLTS.T22695289A174413323.en. Retrieved 19 November 2021.
- Wells, John C. 2008. Longman Pronunciation Dictionary 3rd ed.. Longman. ISBN 978-1-4058-8118-0.
124p - Durant, S.; Mitchell, N.; Ipavec, A. & Groom, R. 2015. "Acinonyx jubatus". IUCN Red List of Threatened Species. 2015: e.T219A50649567. doi:10.2305/IUCN.UK.2015-4.RLTS.T219A50649567.en. Retrieved 15 January 2022.
- Krausman, P. R. & Morales, S. M. 2005. "Acinonyx jubatus" PDF. Mammalian Species. 771: 1–6. doi:10.1644/1545-14102005771[0001:aj]2.0.co;2. S2CID 198969000. Archived from the original PDF on 4 March 2016.
126p - Durant, S.; Mitchell, N.; Ipavec, A. & Groom, R. 2015. "Acinonyx jubatus". IUCN Red List of Threatened Species. 2015: e.T219A50649567. doi:10.2305/IUCN.UK.2015-4.RLTS.T219A50649567.en. Retrieved 15 January 2022.
- Krausman, P. R. & Morales, S. M. 2005. "Acinonyx jubatus" PDF. Mammalian Species. 771: 1–6. doi:10.1644/1545-14102005771[0001:aj]2.0.co;2. S2CID 198969000. Archived from the original PDF on 4 March 2016.
128p - Irwin, M.; King, T.; Ravoloharimanitra, M.; Razafindramanana, J.; Tecot, S. 2021 [amended version of 2020 assessment]. "Eulemur rubriventer". IUCN Red List of Threatened Species. 2021: e.T8203A189740044. Retrieved 6 April 2021.
- Groves, C. P. 2005. Wilson, D. E.; Reeder, D. M. eds.. Mammal Species of the World: A Taxonomic and Geographic Reference 3rd ed.. Baltimore: Johns Hopkins University Press. p. 116. ISBN 0-801-88221-4. OCLC 62265494.
130p - Wacher, T.; Bauman, K. & Cuzin, F. 2015. "Vulpes zerda". IUCN Red List of Threatened Species. 2015: e.T41588A46173447. doi:10.2305/IUCN.UK.2015-4.RLTS.T41588A46173447.en. Retrieved 19 November 2021.
- "fennec". Merriam-Webster. Retrieved 30 January 2020.
134p - Groves, C. P. (2005). Wilson, D. E.; Reeder, D. M. (eds.). Mammal Species of the World: A Taxonomic and Geographic Reference (3rd ed.). Baltimore: Johns Hopkins University Press. p. 157. ISBN 0-801-88221-4. OCLC 62265494.
- Ukizintambara, T.; Olupot, W.; Hart, J. (2019). "Allochrocebus lhoesti". IUCN Red List of Threatened Species. 2019: e.T4220A92345122. doi:10.2305/IUCN.UK.2019-3.RLTS.T4220A92345122.en. Retrieved 12 November 2021.
136p - Wozencraft, W. C. 2005. "Species Panthera leo". In Wilson, D. E.; Reeder, D. M. eds.. Mammal Species of the World: A Taxonomic and Geographic Reference 3rd ed.. Johns Hopkins University Press.
- Bauer, H.; Packer, C.; Funston, P. F.; Henschel, P. & Nowell, K. 2017 [errata version of 2016 assessment]. "Panthera leo". IUCN Red List of Threatened Species. 2016: e.T15951A115130419. doi:10.2305/IUCN.UK.2016-3.RLTS.T15951A107265605.en. Retrieved 15 January 2022.
138p - Jordan, N.R. & Do Linh San, E. 2015. "Suricata suricatta". IUCN Red List of Threatened Species. 2015: e.T41624A45209377. doi:10.2305/IUCN.UK.2015-4.RLTS.T41624A45209377.en. Retrieved 19 November 2021.
- Staaden, M. J. 1994. "Suricata suricatta" PDF. Mammalian Species 483: 1–8. doi:10.2307/3504085. JSTOR 3504085. Archived from the original PDF on 15 March 2016.
140p - Ejigu, D. 2020 [errata version of 2020 assessment]. "Capra walie". IUCN Red List of Threatened Species. 2020: e.T3797A178652661. doi:10.2305/IUCN.UK.2020-2.RLTS.T3797A178652661.en. Retrieved 26 April 2021.
- Berihun 2016. "DNA Metabarcoding Reveals Diet Overlap between the Endangered Walia Ibex and Domestic Goats - Implications for Conservation". PLOS ONE. 11 7: e0159133. Bibcode:2016PLoSO..1159133G. doi:10.1371/journal.pone.0159133. PMC 4945080. PMID 27416020.
142p - Müller, C. 1855–61. Geographi graeci minores. pp. 1.1–14: text and

trans. Ed. J. Blomqvist 1979.
- Hair, P. E. H. 1987. "The Periplus of Hanno in the history and historiography of Black Africa". History in Africa. 14: 43–66. doi:10.2307/3171832. JSTOR 3171832. S2CID 162671887.

144p - Koepfli KP, Deere KA, Slater GJ, et al. 2008. "Multigene phylogeny of the Mustelidae: Resolving relationships, tempo and biogeographic history of a mammalian adaptive radiation". BMC Biol. 6: 4–5. doi:10.1186/1741-7007-6-10. PMC 2276185. PMID 18275614.
- Geraads, Denis; Alemseged, Zeresenay; Bobe, René; Reed, Denné 2011. "Enhydriodon dikikae, sp. nov. Carnivora: Mammalia, a gigantic otter from the Pliocene of Dikika, Lower Awash, Ethiopia". Journal of Vertebrate Paleontology. 31 2: 447–453. doi:10.1080/02724634.2011.550356. S2CID 84797296.

146p - Louis, E.E.; Bailey, C.A.; Sefczek, T.M.; King, T.; Radespiel, U.; Frasier, C.L. 2020. "Propithecus coquereli". IUCN Red List of Threatened Species. 2020: e.T18355A115572275. doi:10.2305/IUCN.UK.2020-2.RLTS.T18355A115572275.en. Retrieved 19 November 2021.
- Groves, C. P. 2005. Wilson, D. E.; Reeder, D. M. eds.. Mammal Species of the World: A Taxonomic and Geographic Reference 3rd ed.. Baltimore: Johns Hopkins University Press. p. 120. ISBN 0-801-88221-4. OCLC 62265494

150p - Groves, C. P. 2005. Wilson, D. E.; Reeder, D. M. eds.. Mammal Species of the World: A Taxonomic and Geographic Reference 3rd ed.. Baltimore: Johns Hopkins University Press. p. 157. ISBN 0-801-88221-4. OCLC 62265494.
- Mwenja, I.; Maisels, F.; Hart, J.A. 2019. "Cercopithecus neglectus". IUCN Red List of Threatened Species. 2019: e.T4223A17947167. doi:10.2305/IUCN.UK.2019-3.RLTS.T4223A17947167.en. Retrieved 19 November 2021.

152p - Louis, E.E.; Bailey, C.A.; Sefczek, T.M.; King, T.; Radespiel, U.; Frasier, C.L. 2020. "Propithecus coquereli". IUCN Red List of Threatened Species. 2020: e.T18355A115572275. doi:10.2305/IUCN.UK.2020-2.RLTS.T18355A115572275.en. Retrieved 19 November 2021.
- Groves, C. P. 2005. Wilson, D. E.; Reeder, D. M. eds.. Mammal Species of the World: A Taxonomic and Geographic Reference 3rd ed.. Baltimore: Johns Hopkins University Press. p. 120. ISBN 0-801-88221-4. OCLC 62265494.

154p - BirdLife International 2017. "Gypaetus barbatus". IUCN Red List of Threatened Species. 2017: e.T22695174A154813652. doi:10.2305/IUCN.UK.2021-3.RLTS.T22695174A154813652.en. Retrieved 19 November 2021.
- Gill F, D Donsker & P Rasmussen Eds. 2021. IOC World Bird List v11.1. doi:10.14344/IOC.ML.11.1

156p - Louis EE, Sefczek TM, Randimbiharinirina DR, Raharivololona B, Rakotondrazandry JN, Manjary D, Aylward M, Ravelomandrato F 2020. "Daubentonia madagascariensis". IUCN Red List of Threatened Species. 2020: e.T6302A115560793. Retrieved 2020-07-18.
- Erickson, C.J.; Nowicki, S.; Dollar, L.; Goehring, N. 1998. "Percussive Foraging: Stimuli for Prey Location by Aye-Ayes Daubentonia madagascariensis". International Journal of Primatology. 19 1: 111. doi:10.1023/A:1020363128240. S2CID 27737088.

158p - King, T.; Dolch, R.; Randriahaingo, H.N.T.; Randrianarimanana, L.; Ravaloharimanitra, M. 2020. "Indri indri". IUCN Red List of Threatened Species. 2020: e.T10826A115565566. doi:10.2305/IUCN.UK.2020-2.RLTS.T10826A115565566.en. Retrieved 19 November 2021.
- Groves, C. P. 2005. Wilson, D. E.; Reeder, D. M. eds.. Mammal Species of the World: A Taxonomic and Geographic Reference 3rd ed.. Baltimore: Johns Hopkins University Press. p. 120. ISBN 0-801-88221-4. OCLC 62265494.

160p - Wozencraft, W. C. 2005. "Genus Profelis". In Wilson, D. E.; Reeder, D. M. eds.. Mammal Species of the World: A Taxonomic and Geographic Reference 3rd ed.. Johns Hopkins University Press. p. 544. ISBN 978-0-8018-8221-0. OCLC 62265494.
- Bahaa-el-din, L.; Mills, D.; Hunter, L. & Henschel, P. 2015. "Caracal aurata". IUCN Red List of Threatened Species. 2015: e.T18306A50663128. doi:10.2305/IUCN.UK.2015-2.RLTS.T18306A50663128.en. Retrieved 25 November 2021.

162p - Groves, C. P. 2005. Wilson, D. E.; Reeder, D. M. eds.. Mammal Species of the World: A Taxonomic and Geographic Reference 3rd ed.. Baltimore: Johns Hopkins University Press. pp. 154–155. ISBN 0-801-88221-4. OCLC 62265494.
- de Jong, Y.A.; Butynski, T.M. 2019. "Cercopithecus ascanius". IUCN Red List of Threatened Species. 2019: e.T4212A17947340. doi:10.2305/IUCN.UK.2019-3.RLTS.T4212A17947340.en. Retrieved 16 November 2021.

166p - Reuter, K.E.; Eppley, T.M.; Hending, D.; Pacifici, M.; Semel, B.; Zaonarivelo, J. 2020. "Eulemur coronatus". IUCN Red List of Threatened Species. 2020: e.T8199A182239524. doi:10.2305/IUCN.UK.2020-3.RLTS.T8199A182239524.en. Retrieved 19 November 2021.
- Groves, C. P. 2005. Wilson, D. E.; Reeder, D. M. eds.. Mammal Species of the World: A Taxonomic and Geographic Reference 3rd ed.. Baltimore: Johns Hopkins University Press. p. 115. ISBN 0-801-88221-4. OCLC 62265494.

168p - IUCN SSC Antelope Specialist Group 2016. "Addax nasomaculatus". IUCN Red List of Threatened Species. 2016: e.T512A50180603. doi:10.2305/IUCN.UK.2016-2.RLTS.T512A50180603.en. Retrieved 3 June 2021.
- Wilson, D.E.; Reeder, D.M., eds. 2005. Mammal Species of the World: A Taxonomic and Geographic Reference 3rd ed.. Johns Hopkins University Press. p. 717. ISBN 978-0-8018-8221-0. OCLC 62265494.

170p - Marino, J.; Sillero-Zubiri, C. 2011. "Canis simensis". IUCN Red List of Threatened Species. 2011: e.T3748A10051312. doi:10.2305/IUCN.UK.2011-1.RLTS.T3748A10051312.en. Retrieved 19 November 2021.
- Wozencraft, C. W. 2005. "Order Carnivora". In Wilson, D. E.; Reader, D. M. eds.. Mammal Species of the World: A Taxonomic and Geographic Reference. Vol. 1 3rd ed.. Johns Hopkins University Press. p. 577. ISBN 978-0-8018-8221-0.

172p - Groves, C. P. 2005. Wilson, D. E.; Reeder, D. M. eds.. Mammal Species of the World: A Taxonomic and Geographic Reference 3rd ed.. Baltimore: Johns Hopkins University Press. p. 167. ISBN 0-801-88221-4. OCLC 62265494.
- Gippoliti, S. & Hunter, C. 2008. "Theropithecus gelada". The IUCN Red List of Threatened Species. 2008: e.T21744A9316114. doi:10.2305/IUCN.UK.2008.RLTS.T21744A9316114.en.

174p - Groves, C. P. (2005). Wilson, D. E.; Reeder, D. M. (eds.). Mammal Species of the World: A Taxonomic and Geographic Reference (3rd ed.). Baltimore: Johns Hopkins University Press. p. 158. ISBN 0-801-88221-4. OCLC 62265494.
- Hart, J.A.; Detwiler, K.M.; Maisels, F. (2020). "Cercopithecus wolfi". IUCN Red List of Threatened Species. 2020: e.T92466239A166601223. doi:10.2305/IUCN.UK.2020-1.RLTS.T92466239A166601223.en. Retrieved 12 November 2021.

176p - Martínez-Navarro, B. & Rook, L. 2003. "Gradual evolution in the African hunting dog lineage: systematic implications". Comptes Rendus Palevol. 2 8: 695–702. doi:10.1016/j.crpv.2003.06.002.
- Woodroffe, R. & Sillero-Zubiri, C. 2020 [amended version of 2012 assessment]. "Lycaon pictus". IUCN Red List of Threatened Species. 2020: e.T12436A166502262. doi:10.2305/IUCN.UK.2020-1.RLTS.T12436A166502262.en. Retrieved 17 February 2022.

178p - "Checklist of CITES Species". CITES. UNEP-WCMC. Retrieved 18 March 2015
- LaFleur, M. & Gould, L. 2020 여우원숭이 카타. IUCN 멸종위기종 목록 2020: e.T11496A115565760. https://dx.doi.org/10.2305/IUCN.UK.2020-

182p - Wozencraft, W. C. (2005). "Species Cryptoprocta ferox". In Wilson, D. E.; Reeder, D. M. (eds.). Mammal Species of the World: A Taxonomic and Geographic Reference (3rd ed.). Johns Hopkins University Press. pp. 559–561. ISBN 978-0-8018-8221-0. OCLC 62265494.
- Hawkins, F. (2016). "Cryptoprocta ferox". IUCN Red List of Threatened Species. 2016: e.T5760A45197189. doi:10.2305/IUCN.UK.2016-1.RLTS.T5760A45197189.en. Retrieved 24 January 2022.

184p - Johnson, S.; Narváez-Torres, P.R.; Holmes, S.M.; Wyman, T.M.; Louis, E.E.; Wright, P. (2020). "Eulemur rufifrons". IUCN Red List of Threatened Species. 2020: e.T136269A115581600. doi:10.2305/IUCN.UK.2020-2.RLTS.T136269A115581600.en. Retrieved 19 November 2021.
- Mittermeier, R.; Ganzhorn, J.; Konstant, W.; Glander, K.; Tattersall, I.; Groves, C.; Rylands, A.; Hapke, A.; Ratsimbazafy, J.; Mayor, M.; Louis, E.; Rumpler, Y.; Schwitzer, C. & Rasoloarison, R. (December 2008). "Lemur Diversity in Madagascar". International Journal of Primatology. 29 (6): 1607–1656. doi:10.1007/s10764-008-9317-y. hdl:10161/6237. S2CID 17614597.

186p - BirdLife International. (2017). Corythaeola cristata (amended version of 2016 assessment). The IUCN Red List of Threatened Species. doi:10.2305/IUCN.UK.2017-1.RLTS.T22688425A111660258.en
- Vieillot, Louis Jean Pierre (1816). Analyse d'une Nouvelle Ornithologie Élémentaire (in French). Paris: Deterville/self. p. 68.

188p - Alexander, G.J.; Tolley, K.A.; Bates, M.F.; Mouton, P.L.F.N. 2018. "Smaug giganteus". IUCN Red List of Threatened Species. 2018: e.T5336A115650269. doi:10.2305/IUCN.UK.2018-2.RLTS.T5336A115650269.en. Retrieved 20 November 2021.
- Mouton, P.le.F.N. 2014. Smaug giganteus Smith, 1844. Pp 209. In: Bates, M.F.,

Branch, W.R., Bauer, A.M., Burger, M., Marais, J., Alexander, G.J., De Villiers, M.S. eds.. Atlas and Red List of the Reptiles of South Africa, Lesotho and Swaziland. Suricata 1. South African National Biodiversity Institute, Pretoria.

190p - BirdLife International 2019. "Neophron percnopterus". IUCN Red List of Threatened Species. 2019: e.T22695180A154895845. doi:10.2305/IUCN.UK.2019-3.RLTS.T22695180A154895845.en. Retrieved 11 November 2021.
- Gill F, D Donsker & P Rasmussen Eds. 2020. IOC World Bird List v10.2. doi:10.14344/IOC.ML.10.2.

192p - Hegyeli, Z. 2020. "Spermophilus citellus". IUCN Red List of Threatened Species. 2020: e.T20472A91282380. doi:10.2305/IUCN.UK.2020-2.RLTS.T20472A91282380.en. Retrieved 19 November 2021.
- Thorington, R.W., Jr.; Hoffman, R.S. 2005. "Family Sciuridae". In Wilson, D.E.; Reeder, D.M eds.. Mammal Species of the World: A Taxonomic and Geographic Reference 3rd ed.. Johns Hopkins University Press. p. 805. ISBN 978-0-8018-8221-0. OCLC 62265494.

194p - IUCN SSC Antelope Specialist Group 2016. "Litocranius walleri". IUCN Red List of Threatened Species. 2016: e.T12142A50190292. doi:10.2305/IUCN.UK.2016-2.RLTS.T12142A50190292.en. Retrieved 12 November 2021.
- IUCN SSC Antelope Specialist Group 2008. "Litocranius walleri". IUCN Red List of Threatened Species. 2008. Retrieved 21 June 2012.

196p -wikipedia, "American_bison", https://en.wikipedia.org/wiki/, 2023.04.17/
--gousa, "American_bison",
https://www.gousa.or.kr/experience/where-bison-still-roam-usa

198p - Groves, C. P. (2005). Wilson, D. E.; Reeder, D. M. (eds.). Mammal Species of the World: A Taxonomic and Geographic Reference (3rd ed.). Baltimore: Johns Hopkins University Press. p. 168. ISBN 0-801-88221-4. OCLC 62265494.
- de Jong, Y.A.; Cunneyworth, P; Butynski, T.M.; Maisels, F.; Hart, J.A.; Rovero, F. (2020). "Colobus angolensis". IUCN Red List of Threatened Species. 2020: e.T5142A17945007. doi:10.2305/IUCN.UK.2020-2.RLTS.T5142A17945007.en. Retrieved 19 November 2021.

200p - Groves, C. P. (2005). Wilson, D. E.; Reeder, D. M. (eds.). Mammal Species of the World: A Taxonomic and Geographic Reference (3rd ed.). Baltimore: Johns Hopkins University Press. p. 155. ISBN 0-801-88221-4. OCLC 62265494.
- Koné, I.; McGraw, S.; Gonedelé Bi, S.; Oates, J.F. (2019). "Cercopithecus diana". IUCN Red List of Threatened Species. 2019: e.T4245A92384250. doi:10.2305/IUCN.UK.2019-3.RLTS.T4245A92384250.en. Retrieved 19 November 2021.

202p - Groves, C. P. (2005). Wilson, D. E.; Reeder, D. M. (eds.). Mammal Species of the World: A Taxonomic and Geographic Reference (3rd ed.). Baltimore: Johns Hopkins University Press. p. 169. ISBN 0-801-88221-4. OCLC 62265494.
- McGraw, S.; Minhós, T.; Bersacola, E.; Ferreira da Silva, M.J.; Galat, G.; Galat-Luong, A.; Gonedelé Bi, S.; Mayhew, M.; Oates, J.F.; Starin, E.D. (2020). "Piliocolobus badius". IUCN Red List of Threatened Species. 2020: e.T161247840A161259430. doi:10.2305/IUCN.UK.2020-1.RLTS.T161247840A161259430.en. Retrieved 19 November 2021.

204p - Tedford, Richard H.; Wang, Xiaoming; Taylor, Beryl E. 2009. "Phylogenetic Systematics of the North American Fossil Caninae Carnivora: Canidae". Bulletin of the American Museum of Natural History. 325: 1–218. doi:10.1206/574.1. hdl:2246/5999. S2CID 83594819.
- Boitani, L.; Phillips, M.; Jhala, Y. 2018. "Canis lupus". IUCN Red List of Threatened Species. 2018: e.T3746A163508960. doi:10.2305/IUCN.UK.2018-2.RLTS.T3746A163508960.en. Retrieved November 19, 2021.

206p - Groves, C. P. (2005). Wilson, D. E.; Reeder, D. M. (eds.). Mammal Species of the World: A Taxonomic and Geographic Reference (3rd ed.). Baltimore: Johns Hopkins University Press. p. 155. ISBN 0-801-88221-4. OCLC 62265494.
- Abernethy, K.; Maisels, F. (2020). "Cercopithecus cephus". IUCN Red List of Threatened Species. 2020: e.T4214A166614362. doi:10.2305/IUCN.UK.2020-1.RLTS.T4214A166614362.en. Retrieved 17 November 2021.

208p - Groves, C. P. (2005). Wilson, D. E.; Reeder, D. M. (eds.). Mammal Species of the World: A Taxonomic and Geographic Reference (3rd ed.). Baltimore: Johns Hopkins University Press. p. 159. ISBN 0-801-88221-4. OCLC 62265494.
- Gippoliti, S.; Butynski, T.M.; Mekonnen, A. (2019). "Chlorocebus djamdjamensis". IUCN Red List of Threatened Species. 2019: e.T4240A17958005. doi:10.2305/IUCN.UK.2019-3.RLTS.T4240A17958005.en. Retrieved 19 November 2021.

210p - IUCN SSC Antelope Specialist Group. 2017 [errata version of 2016 assessment]. "Tragelaphus eurycerus". IUCN Red List of Threatened Species. 2016: e.T22047A115164600. Retrieved 24 October 2020.
- IUCN SSC Antelope Specialist Group. 2017. "Tragelaphus eurycerus ssp. eurycerus". IUCN Red List of Threatened Species. 2017: e.T22058A50197275. doi:10.2305/IUCN.UK.2017-2.RLTS.T22058A50197275.en. Retrieved 24 October 2020.

212p -wikipedia, https://en.wikipedia.org, "wiki/Arabian_oryx", 2023.04.17
- 애니멀피플,「누가 아라비아 오릭스를 되살렸을까?」, https://www.hani.co.kr/arti/animalpeople/wild_animal/994490.html, 2023.04.17

214p - Avgan, B.; Henschel, P & Ghoddousi, A. 2016 [errata version of 2016 assessment]. "Caracal caracal". IUCN Red List of Threatened Species. 2016: e.T3847A102424310. doi:10.2305/IUCN.UK.2016-
- Wozencraft, W. C. 2005. "Species Carcal caracal". In Wilson, D. E.; Reeder, D. M. eds.. Mammal Species of the World: A Taxonomic and Geographic Reference 3rd ed.. Johns Hopkins University Press. p. 533. ISBN 978-0-8018-8221-0. OCLC 62265494.

216p - West Indian Manatee Facts and Pictures – National Geographic Kids Archived 2011-06-26 at the Wayback Machine. Kids.nationalgeographic.com. Retrieved on 2011-12-03.
- Winger, Jennifer (2000). "What's in a name? Manatees and Dugongs". National Zoological Park. Friends of the National Zoo. Archived from the original on 30 December 2005. Retrieved 19 June 2015.

218p - Groves, C. P. 2005. "Chlorocebus pygerythrus". In Wilson, D. E.; Reeder, D. M. eds.. Mammal Species of the World: A Taxonomic and Geographic Reference 3rd ed.. Baltimore: Johns Hopkins University Press. p. 159. ISBN 0-801-88221-4. OCLC 62265494.
- Butynski, T.M.; de Jong, Y.A. 2019. "Chlorocebus pygerythrus". IUCN Red List of Threatened Species. 2019: e.T136271A17957823. doi:10.2305/IUCN.UK.2019-3.RLTS.T136271A17957823.en. Retrieved 19 November 2021.

220p - AbiSaid, M.; Dloniak, S.M.D. (2015). "Hyaena hyaena". IUCN Red List of Threatened Species. 2015: e.T10274A45195080. doi:10.2305/IUCN.UK.2015-2.RLTS.T10274A45195080.en. Retrieved 19 November 2021.
- Linnæus, C. (1758). Systema naturæ per regna tria naturæ, secundum classes, ordines, genera, species, cum characteribus, differentiis, synonymis, locis. Tomus I (in Latin) (Tenth ed.). Holmiæ (Stockholm): Laurentius Salvius. p. 40.

222p - BirdLife International 2017. "Bubo bubo". IUCN Red List of Threatened Species. 2017: e.T22688927A113569670. doi:10.2305/IUCN.UK.2017-1.RLTS.T22688927A113569670.en. Retrieved 19 November 2021.
- "Eurasian Eagle-Owl Bubo bubo Linnaeus, 1758". AviBase. Retrieved 2014-07-25.

224p - BirdLife International (2018). "Fratercula arctica". IUCN Red List of Threatened Species. 2018: e.T22694927A132581443. doi:10.2305/IUCN.UK.2018-2.RLTS.T22694927A132581443.en. Retrieved 19 November 2021.
- Smith, N.A. (2011). "Taxonomic revision and phylogenetic analysis of the flightless Mancallinae (Aves, Pan-Alcidae)". ZooKeys (91): 1–116. doi:10.3897/zookeys.91.709. PMC 3084493. PMID 21594108.

226p -wikipedia, "Bald_eagle ", https://en.wikipedia.org/wiki/, 2023.04.17
-BRIC Bio 통신원, 미국 대머리독수리, 멸종위기종 목록에서 제외, https://www.ibric.org/myboard/read.php?Board=news&id=112072

228p - BirdLife International 2017. "Gypaetus barbatus". IUCN Red List of Threatened Species. 2017: e.T22695174A154813652. doi:10.2305/IUCN.UK.2021-3.RLTS.T22695174A154813652.en. Retrieved 19 November 2021.
- Gill F, D Donsker & P Rasmussen Eds. 2021. IOC World Bird List v11.1. doi:10.14344/IOC.ML.11.1

230p - Gorsuch, W. & Larivière, S. (2005). "Vormela peregusna". Mammalian Species. 779: 1–5. doi:10.1644/779.1.
- Stroganov, S.U. (1969). Carnivorous mammals of Siberia. Jerusalem: Israeli Program of Scientific Translation. ISBN 0-7065-0645-6.

232p - Groves, C. P. 2005. "Chlorocebus cynosuros". In Wilson, D. E.; Reeder, D. M. eds.. Mammal Species of the World: A Taxonomic and Geographic Reference 3rd ed.. Baltimore: Johns Hopkins University Press. p. 159. ISBN 0-801-88221-4. OCLC 62265494.
- Wallis, J. 2019. "Chlorocebus cynosuros". IUCN Red List of Threatened Species. 2019: e.T136291A17957592. doi:10.2305/IUCN.UK.2019-3.RLTS.T136291A17957592.en. Retrieved 19 November 2021.

236p - Woinarski, J.; Burbidge, A.A. 2020. "Phascolarctos cinereus". IUCN Red List of Threatened Species. 2020: e.T16892A166496779. doi:10.2305/IUCN.UK.2020-1.RLTS.T16892A166496779.en. Retrieved 12 November 2021.
- Moyal, Ann 2008. Koala: a historical biography. Melbourne: CSIRO Pub. ISBN 978-0-643-09401-7. OCLC 476194354. Archived from the original on 2 May 2016. Retrieved 9 November 2015.

238p - Shar, S.; Lkhagvasuren, D.; Henttonen, H.; Maran, T. & Hanski, I. (2017) [errata version of 2016 assessment]. "Pteromys volans". IUCN Red List of Threatened Species. 2016: e.T18702A115144995. doi:10.2305/IUCN.UK.2016-3.RLTS.T18702A22270935.en. Retrieved 17 February 2022.
- Ikauniece, S. "Lidvāveres atstājušas Latviju". Dabas aizsardzības pārvalde (in Latvian). Retrieved 6 January 2023.

240p - Woinarski, J.; Burbidge, A.A. 2016. "Ornithorhynchus anatinus". IUCN Red List of Threatened Species. 2016: e.T40488A21964009. doi:10.2305/IUCN.UK.2016-1.RLTS.T40488A21964009.en. Retrieved 19 November 2021.
- "Ornithorhynchus anatinus". Global Biodiversity Information Facility. Retrieved 13 July 2021.

242p - Burbidge, A.A.; Johnson, C.N.; Zichy-Woinarski, J. 2016. "Onychogalea fraenata". IUCN Red List of Threatened Species. 2016: e.T15330A21958130. doi:10.2305/IUCN.UK.2016-1.RLTS.T15330A21958130.en. Retrieved 12 November 2021.
- Gould, J. 1841. "On five new species of kangaroo". Proceedings of the Zoological Society of London. 1840: 92–94.

244p - Groves, C. P. (2005). Wilson, D. E.; Reeder, D. M. (eds.). Mammal Species of the World: A Taxonomic and Geographic Reference (3rd ed.). Baltimore: Johns Hopkins University Press. p. 51. ISBN 0-801-88221-4. OCLC 62265494.
- Winter, J.; Menkhorst, P.; Lunney, D.; van Weenen, J. (2016). "Pseudocheirus peregrinus". IUCN Red List of Threatened Species. 2016: e.T40581A21963019. doi:10.2305/IUCN.UK.2016-2.RLTS.T40581A21963019.en. Retrieved 12 November 2021.

246p - Groves, C. P. (2005). Wilson, D. E.; Reeder, D. M. (eds.). Mammal Species of the World: A Taxonomic and Geographic Reference (3rd ed.). Baltimore: Johns Hopkins University Press. p. 51. ISBN 0-801-88221-4. OCLC 62265494.
- McGregor, Denise C.; Padovan, Amanda; Georges, Arthur; Krockenberger, Andrew; Yoon, Hwan-Jin; Youngentob, Kara N. (2020). "Genetic evidence supports three previously described species of greater glider, Petauroides volans, P. minor, and P. armillatus". Scientific Reports. 10 (1): 19284.

248p -wikipedia, "Brown_pelican ", https://en.wikipedia.org/wiki/, 2023.04.17 -Halle Parker, "brown-pelican", https://www.audubon.org/news/finding-hope-story-brown-pelican

250p Reid, J. (1837). "Description of a new species of the genus Perameles (P. lagotis)". Proceedings of the Zoological Society of London. 1836: 129–131. Archived from the original on 14 July 2021. Retrieved 27 July 2019.
- Wilson & Reeder. "Genus Macrotis". Mammal Species of the World. Archived from the original on 27 August 2017. Retrieved 14 August 2014.

252p - Burbidge, A.A. & Woinarski, J. 2020 [amended version of 2019 assessment]. "Setonix brachyurus". IUCN Red List of Threatened Species. 2020: e.T20165A166611530. doi:10.2305/IUCN.UK.2020-1.RLTS.T20165A166611530.en. Retrieved 7 April 2021.
- Quoy, [Jean René Constant]; Gaimard, [Joseph Paul] 1830. "Kangurus brachyurus". Voyage de découvertes de l'Astrolabe: Zoologie. Vol. 1. Paris: J. Tastu. pp. 114–116.

254p - Groves, C. P. 2005. "Order Primates". In Wilson, D. E.; Reeder, D. M eds.. Mammal Species of the World: A Taxonomic and Geographic Reference 3rd ed.. Johns Hopkins University Press. p. 180. ISBN 978-0-8018-8221-0. OCLC 62265494.
- Pengfei, F.; Nguyen, M.H.; Phiaphalath, P.; Roos, C.; Coudrat, C.N.Z.; Rawson, B.M. 2020. "Nomascus concolor". IUCN Red List of Threatened Species. 2020: e.T39775A17968556. doi:10.2305/IUCN.UK.2020-2.RLTS.T39775A17968556.en. Retrieved 19 November 2021.

256p - McCarthy, T.; Mallon, D.; Jackson, R.; Zahler, P. & McCarthy, K. 2017. "Panthera uncia". IUCN Red List of Threatened Species. 2017: e.T22732A50664030. doi:10.2305/IUCN.UK.2017-2.RLTS.T22732A50664030.en. Retrieved 19 November 2021.
- Allen, E.A. 1908. "English Doublets". Publications of the Modern Language Association of America. New Series 16. 23 1: 184–239. doi:10.2307/456687.

JSTOR 456687. S2CID 251028590.

258p - Wilson, D.E.; Reeder, D.M., eds. 2005. Mammal Species of the World: A Taxonomic and Geographic Reference 3rd ed.. Johns Hopkins University Press. ISBN 978-0-8018-8221-0. OCLC 62265494.
- "Checklist of CITES Species". CITES. UNEP-WCMC. Retrieved 18 March 2015.

260p - Abramov, A.V.; Duckworth, J.W.; Choudhury, A.; Chutipong, W.; Timmins, R.J.; Ghimirey, Y.; Chan, B.; Dinets, V. 2016. "Mustela sibirica". IUCN Red List of Threatened Species. 2016: e.T41659A45214744. doi:10.2305/IUCN.UK.2016-1.RLTS.T41659A45214744.en. Retrieved 19 November 2021.
- Heptner & Sludskii 2002, pp. 1052–1054

262p - Rak, K. C.; Miquelle, D. G. & Pikunov, D. G. 1998. A survey of tigers and leopards and prey resources in the Paektusan area, North Korea, in winter 1998 PDF Report. Archived from the original on 5 December 2020. Retrieved 16 June 2012.
- Goodrich, J.; Lynam, A.; Miquelle, D.; Wibisono, H.; Kawanishi, K.; Pattanavibool, A.; Htun, S.; Tempa, T.; Karki, J.; Jhala, Y. & Karanth, U. 2015. "Panthera tigris". IUCN Red List of Threatened Species. 2015: e.T15955A50659951
- WWF Russia 2015. "Russia Announce Tiger Census Results!". tigers.panda.org. Worldwide Fund for Nature. Retrieved June 7, 2015.

264p -wikipedia, "Humpback_whale ", https://en.wikipedia.org/wiki/, 2023.04.17 /-우성희, 멸종위기생물종 리스트에서 제외될 혹등고래, http://mire.re.kr/sub4_3.php?type=read&code=board5&id=662047&page=4&part=&word=&domain=&PHPSESSID=793eaa28b70b4694e96d753e927febfc, 2023.04.17

266p - Groves, C. P. 2005. "Order Primates". In Wilson, D. E.; Reeder, D. M eds.. Mammal Species of the World: A Taxonomic and Geographic Reference 3rd ed.. Johns Hopkins University Press. pp. 168–169. ISBN 978-0-8018-8221-0. OCLC 62265494.
- Boonratana, R.; Cheyne, S.M.; Traeholt, C.; Nijman, V. & Supriatna, J. 2021. "Nasalis larvatus". IUCN Red List of Threatened Species. 2021: e.T14352A195372486. doi:10.2305/IUCN.UK.2021-1.RLTS.T14352A195372486.en. Retrieved 17 January 2022.

268p - Groves, C. P. (2005). Wilson, D. E.; Reeder, D. M. (eds.). Mammal Species of the World: A Taxonomic and Geographic Reference (3rd ed.). Baltimore: Johns Hopkins University Press. p. 55. ISBN 0-801-88221-4. OCLC 62265494.
- Salas, L.; Dickman, C.; Helgen, K.; Winter, J.; Ellis, M.; Denny, M.; Woinarski, J.; Lunney, D.; Oakwood, M.; Menkhorst, P.; Strahan, R. (2016). "Petaurus breviceps". IUCN Red List of Threatened Species. 2016. e.T16731A21959798. doi:10.2305/IUCN.UK.2016-2.RLTS.T16731A21959798.en. Retrieved 19 November 2021.

270p - Abramov, A.V.; Kaneko, Y.; Masuda, R. 2015. "Martes melampus". IUCN Red List of Threatened Species. 2015: e.T41650A45213228. doi:10.2305/IUCN.UK.2015-4.RLTS.T41650A45213228.en. Retrieved 25 September 2021.
- Buskirk, Steven September 1992. "Conserving Circumboreal Forests for Martens and Fishers". Conservation Biology. 6 3: 318–323. doi:10.1046/j.1523-1739.1992.06030318.x.

272p - Watanabe, K.; Tokita, K. 2020. "Macaca fuscata". IUCN Red List of Threatened Species. 2020: e.T12552A195347803. doi:10.2305/IUCN.UK.2020-2.RLTS.T12552A195347803.en. Retrieved 11 November 2021.
- Blyth, Edward 1875. "Catalogue of the Mammals and Birds of Burma". The Journal of the Asiatic Society of Bengal. 44 1? Extra Number: 6.

274p - Swaisgood, R.; Wang, D. & Wei, F. 2017 [errata version of 2016 assessment]. "Ailuropoda melanoleuca". IUCN Red List of Threatened Species. 2016: e.T712A121745669. Retrieved 15 January 2022.
- David, A. 1869. "Voyage en Chine". Bulletin des Nouvelles Archives du Muséum. 5: 13.

276p - BirdLife International 2016. "Strix aluco". IUCN Red List of Threatened Species. 2016: e.T22725469A86871093. doi:10.2305/IUCN.UK.2016-3.RLTS.T22725469A86871093.en. Retrieved 19 November 2021.
- Based on Güntürkün, Onur; "Structure and functions of the eye" in Sturkie, P. D. 1998. Sturkie's Avian Physiology. 5th Edition. Academic Press, San Diego. pp. 1–18. ISBN 978-0-12-747605-6.

278p - Shekelle, M.; Salim, A. 2020. "Tarsius tumpara". IUCN Red List of Threatened Species. 2020: e.T179234A17977202. doi:10.2305/IUCN.UK.2020-3.RLTS.T179234A17977202.en. Retrieved 12 November 2021.
- Shekelle, M.; Groves, C.; Merker, S.; Supriatna, J. 2008. "Tarsius tumpara: A new tarsier species from Siau Island, North Sulawesi" PDF. Primate Conservation.

23: 55–64. doi:10.1896/052.023.0106. S2CID 55493260.
280p -wikipedia, "wiki/Kihansi_spray_toad", ttps://en.wikipedia.org/wiki/, 2023.04.17
282p - Glatston, A.; Wei, F.; Than Zaw & Sherpa, A. 2017 [errata version of 2015 assessment]. "Ailurus fulgens". IUCN Red List of Threatened Species. 2015: e.T714A110023718. Retrieved 15 January 2022.
- Thomas, O. 1902. "On the Panda of Sze-chuen". Annals and Magazine of Natural History. 7. X 57: 251–252. doi:10.1080/00222930208678667.
284p - Garshelis, D. & Steinmetz, R. 2020. "Ursus thibetanus". IUCN Red List of Threatened Species. 2020: e.T22824A166528664. doi:10.2305/IUCN.UK.2020-3.RLTS.T22824A166528664.en. Retrieved 16 January 2022.
- Montgomery, S. 2002. Search for the golden moon bear: science and adventure in Southeast Asia. Simon & Schuster. ISBN 978-0-7432-0584-9.
286p - Groves, C. P. 2005. Wilson, D. E.; Reeder, D. M. eds.. Mammal Species of the World: A Taxonomic and Geographic Reference 3rd ed.. Baltimore: Johns Hopkins University Press. p. 120. ISBN 0-801-88221-4. OCLC 62265494.
- Louis, E.E.; Bailey, C.A.; Sefczek, T.M.; King, T.; Radespiel, U.; Frasier, C.L. 2020. "Propithecus coquereli". IUCN Red List of Threatened Species. 2020: e.T18355A115572275. doi:10.2305/IUCN.UK.2020-2.RLTS.T18355A115572275.en. Retrieved 19 November 2021.
288p - Abe, Hisashi, ed. (2008). 日本の哺乳類 (Nihon no Honyuurui) [A Guide to the Mammals of Japan] (in Japanese). Tokai University. ISBN 978-4-486-01802-5.
- Grubb, P. (2005). "Order Artiodactyla". In Wilson, D.E.; Reeder, D.M (eds.). Mammal Species of the World: A Taxonomic and Geographic Reference (3rd ed.). Johns Hopkins University Press. p. 704. ISBN 978-0-8018-8221-0. OCLC 62265494.
290p - Groves, C. P. 2005. Wilson, D. E.; Reeder, D. M. eds.. Mammal Species of the World: A Taxonomic and Geographic Reference 3rd ed.. Baltimore: Johns Hopkins University Press. p. 60. ISBN 0-801-88221-4. OCLC 62265494.
- Ziembicki, M.; Porolak, G. 2016. "Dendrolagus matschiei". IUCN Red List of Threatened Species. 2016: e.T6433A21956650. doi:10.2305/IUCN.UK.2016-2.RLTS.T6433A21956650.en. Retrieved 19 November 2021.
292p - J. H. Schwartz, T. H.Vu, L. C. Nguyen, K.T. Le, and I. Tattersall. 1994. A diverse hominoid fauna from the late middle Pleistocene breccia cave of Tham Khuyen, Socialist Republic of Vietnam. Anthropological Papers of the American Museum of Natural History 73:1-11
- Scotson, L.; Fredriksson, G.; Augeri, D.; Cheah, C.; Ngoprasert, D. & Wai-Ming, W. 2018 [errata version of 2017 assessment]. "Helarctos malayanus". IUCN Red List of Threatened Species. 2017: e.T9760A123798233. Retrieved 16 January 2022.
294p - Duckworth, J.W.; Mathai, J.; Wilting, A.; Holden, J.; Hearn, A.; Ross, J. 2016. "Viverra tangalunga". IUCN Red List of Threatened Species. 2016: e.T41708A45220284. doi:10.2305/IUCN.UK.2016-1.RLTS.T41708A45220284.en. Retrieved 19 November 2021.
- Gray, J. E. 1832. "On the family of Viverridae and its generic sub-divisions, with an enumeration of the species of several new ones". Proceedings of the Committee of Science and Correspondence of the Zoological Society of London. 2: 63–68.
296p -wikipedia, "Burmese_star_tortoise", https://en.wikipedia.org/wiki/, 2023.04.17/ -Lindsey Jean Schueman, Burmese star tortoise: the stunning conservation success story in Myanmar, https://www.oneearth.org/species-of-the-week-burmese-star-tortoise/
298p - Molur, S. 2016. "Ratufa indica". IUCN Red List of Threatened Species. 2016: e.T19378A22262028. doi:10.2305/IUCN.UK.2016-2.RLTS.T19378A22262028.en. Retrieved 19 November 2021.
- Thorington, R.W. Jr.; Hoffmann, R.S. 2005. "Ratufa indica". In Wilson, D.E.; Reeder, D.M eds.. Mammal Species of the World: A Taxonomic and Geographic Reference 3rd ed.. The Johns Hopkins University Press. pp. 754–818. ISBN 0-8018-8221-4. OCLC 26158608.
300p - Suraprasit, K.; Jaegar, J.-J.; Chaimanee, Y.; Chavasseau, O.; Yamee, C.; Tian, P. & Panha, S. 2016. "The Middle Pleistocene vertebrate fauna from Khok Sung Nakhon Ratchasima, Thailand: biochronological and paleobiogeographical implications". ZooKeys 613: 1–157. doi:10.3897/zookeys.613.8309. PMC 5027644. PMID 27667928.
- Duckworth, J.W.; Sankar, K.; Williams, A.C.; Samba Kumar, N. & Timmins, R.J. 2016. "Bos gaurus". IUCN Red List of Threatened Species. 2016: e.T2891A46363646. doi:10.2305/IUCN.UK.2016-2.RLTS.T2891A46363646.en. Retrieved 15 January 2022.

302p - Wibisono, H.; Wilianto, E.; Pinondang, I.; Rahman, D.A. & Chandradewi, D. 2021. "Panthera pardus ssp. melas". IUCN Red List of Threatened Species. 2021: e.T15962A50660931.
- Cuvier, G. 1809. "Recherches sur les espèces vivantes de grands chats, pour servir de preuves et d'éclaircissement au chapitre sur les carnassiers fossils". Annales du Muséum National d'Histoire Naturelle. Tome XIV: 136–164.
304p - Groves, C. P. 2005. Wilson, D. E.; Reeder, D. M. eds.. Mammal Species of the World: A Taxonomic and Geographic Reference 3rd ed.. Baltimore: Johns Hopkins University Press. p. 172. ISBN 0-801-88221-4. OCLC 62265494.
- Nijman, V. & Lammertink, M. 2008. "Presbytis natunae". The IUCN Red List of Threatened Species. IUCN. 2008: e.T136500A4301419. doi:10.2305/IUCN.UK.2008.RLTS.T136500A4301419.en.
306p - Singleton, Ian; Wich, Serge A.; Nowak, Matthew G.; Usher, Graham; Utami-Atmoko, Sri Suchi 2018 [errata version of 2017 assessment]. "Pongo abelii". IUCN Red List of Threatened Species. 2017: e.T121097935A123797627.
- Lesson, René-Primevère 1827. Manuel de mammalogie ou Histoire naturelle des mammifères in French. Paris: Roret, Libraire. p. 32.
308p - Timmins, R.J.; Duckworth, J.W.; Hedges, S. (2016). "Muntiacus muntjak". IUCN Red List of Threatened Species. 2016: e.T42190A56005589. doi:10.2305/IUCN.UK.2016-1.RLTS.T42190A56005589.en. Retrieved 19 November 2021.
- "Barking Deer". iloveindia.com. Retrieved 9 September 2022. Muntjac deer fall in the category of those deer that are shy and elusive. They are also known by the name of Kakad deer or the Barking deer in India. The reason for the latter name is their alarm call, which seems very much similar to the barking of a dog.
310p - Li, W.; Smith, A.T. (2019). "Ochotona iliensis". IUCN Red List of Threatened Species. 2019: e.T15050A45179204. doi:10.2305/IUCN.UK.2019-1.RLTS.T15050A45179204.en. Retrieved 17 November 2021.
- Li, W.; Smith, A.T. (2005). "Dramatic decline of the threatened Ili pika Ochotona iliensis (Lagomorpha: Ochotonidae) in Xinjiang, China". Oryx. 39: 30–34. doi:10.1017/s0030605305000062.
312p -한국교육포털,「멸종위기 고라니, 한국에선 유해야생동물?」, https://keep.go.kr/portal/141?action=read&action-value=ccc9dd699f5d3afe0f44dec81b87a50a&page=1, 2023.04.17
-애니멀피플,「농사나 축내는 고라니를 왜 구조하냐고요?」, https://www.hani.co.kr/arti/animalpeople/human_animal/814676.html, 2023.04.17
314p - BirdLife International (2017). "Nisaetus bartelsi". IUCN Red List of Threatened Species. 2017: e.T22696165A110050373. doi:10.2305/IUCN.UK.2017-1.RLTS.T22696165A110050373.en. Retrieved 13 November 2021.
- Helbig AJ, Kocum A, Seibold I & Braun MJ (2005)
316p - Harris, R.B.; Duckworth, J.W. 2015. "Hydropotes inermis". IUCN Red List of Threatened Species. 2015: e.T10329A22163569. doi:10.2305/IUCN.UK.2015-2.RLTS.T10329A22163569.en. Retrieved 12 November 2021.
- Grubb, P. 2005. "Order Artiodactyla". In Wilson, D.E.; Reeder, D.M eds.. Mammal Species of the World: A Taxonomic and Geographic Reference 3rd ed.. Johns Hopkins University Press. p. 671. ISBN 978-0-8018-8221-0. OCLC 62265494.
318p - Smith, A.T. and Liu, S. (2019). "Ochotona curzoniae". IUCN Red List of Threatened Species. 2019: e.T41258A160699229. doi:10.2305/IUCN.UK.2019-3.RLTS.T41258A160699229.en. Retrieved 12 November 2021.
- Xu Aichun, Jiang Zhigang, Li Chunwang, Guo Jixun, Wu Guosheng, Cai Ping, "Summer Food Habits of Brown Bears in Kekexili Nature Reserve, Qinghai: Tibetan Plateau, China". Ursus, Vol. 17, No. 2 (2006), pp. 132–137
320p - Boyd, L.; King, S. R. B.; Zimmermann, W. & Kendall, B.E. 2015. "Equus ferus". IUCN Red List of Threatened Species. 2015. Retrieved 16 December 2020.
- "Przewalski's horse". Lexico UK English Dictionary. Oxford University Press.
322p - Wozencraft, W. C. 2005. "Species Arctictis binturong". In Wilson, D. E.; Reeder, D. M. eds.. Mammal Species of the World: A Taxonomic and Geographic Reference 3rd ed.. Johns Hopkins University Press. p. 549. ISBN 978-0-8018-8221-0. OCLC 62265494.
- Willcox, D.H.A.; Chutipong, W.; Gray, T.N.E.; Cheyne, S.; Semiadi, G.; Rahman, H.; Coudrat, C.N.Z.; Jennings, A.; Ghimirey, Y.; Ross, J.; Fredriksson, G.; Tilker, A. 2016. "Arctictis binturong". IUCN Red List of Threatened Species. 2016: e.T41690A45217088. doi:10.2305/IUCN.UK.2016-1.RLTS.T41690A45217088.en. Retrieved 20 November 2021.

324p - BirdLife International 2020. "Megadyptes antipodes". IUCN Red List of Threatened Species. 2020: e.T22697800A182703046.
doi:10.2305/IUCN.UK.2020-3.RLTS.T22697800A182703046.en. Retrieved 12 November 2021.
- Mattern T, Meyer S, Ellenberg U, Houston DM, Darby JD, Young M, van Heezilk Y, Seddon PJ 2017. "Quantifying climate change impacts emphasises the importance of managing regional threats in the endangered Yellow-eyed penguin". PeerJ. 5: e3272. doi:10.7717/peerj.3272. PMC 5436559. PMID 28533952.

326p -농림축산부검역본부,
https://www.qia.go.kr/animal/prevent/ani_africa_pig_fever.jsp, 2023.04.17
-Vittorio Guberti,Sergei Khomenko,Marius Masiulis,Suzanne Kerba, "멧돼지의 아프리카돼지열병 생태와 차단방역", 국립생태원(2020.04)

328p - BirdLife International (2016). "Spilornis rufipectus". IUCN Red List of Threatened Species. 2016: e.T22695313A93502324.
doi:10.2305/IUCN.UK.2016-3.RLTS.T22695313A93502324.en. Retrieved 13 November 2021.
- Gill F, D Donsker & P Rasmussen (Eds). 2020. IOC World Bird List (v10.2). doi : 10.14344/IOC.ML.10.2.

330p - Kitchener, A. C.; Breitenmoser-Würsten, C.; Eizirik, E.; Gentry, A.; Werdelin, L.; Wilting, A.; Yamaguchi, N.; Abramov, A. V.; Christiansen, P.; Driscoll, C.; Duckworth, J. W.; Johnson, W.; Luo, S.-J.; Meijaard, E.; O'Donoghue, P.; Sanderson, J.; Seymour, K.; Bruford, M.; Groves, C.; Hoffmann, M.; Nowell, K.; Timmons, Z. & Tobe, S. 2017. "A revised taxonomy of the Felidae: The final report of the Cat Classification Task Force of the IUCN Cat Specialist Group" PDF. Cat News Special Issue 11: 21–22.
- Ross, S.; Barashkova, A.; Dhendup, T.; Munkhtsog, B.; Smelansky, I.; Barclay, D. & Moqanaki, E. 2020. "Otocolobus manul". IUCN Red List of Threatened Species. 2020: e.T15640A180145377.
doi:10.2305/IUCN.UK.2020-2.RLTS.T15640A180145377.en. Retrieved 20 November 2021.

332p - Hocknull SA, Piper PJ, van den Bergh GD, Due RA, Morwood MJ, Kurniawan I (2009). "Dragon's Paradise Lost: Palaeobiogeography, Evolution and Extinction of the Largest-Ever Terrestrial Lizards (Varanidae)". PLOS ONE. 4 (9): e7241. Bibcode:2009PLoSO...4.7241H.
doi:10.1371/journal.pone.0007241. PMC 2748693. PMID 19789642.
- Jessop, Tim; Ariefiandy, Achmad; Azmi, Muhammad; Ciofi, Claudio; Imansyah, Jeri; Purwandana, Deni (5 August 2021). "Varanus komodoensis". IUCN Red List of Threatened Species. 2021: e.T22884A123633058.
doi:10.2305/IUCN.UK.2021-2.RLTS.T22884A123633058.en. Retrieved 19 November 2021.

334p - Wozencraft, W. C. (2005). "Species Prionailurus planiceps". In Wilson, D. E.; Reeder, D. M. (eds.). Mammal Species of the World: A Taxonomic and Geographic Reference (3rd ed.). Johns Hopkins University Press. p. 543. ISBN 978-0-8018-8221-0. OCLC 62265494.
- Wilting, A.; Brodie, J.; Cheyne, S.; Hearn, A.; Lynam, A.; Mathai, J.; McCarthy, J.; Meijaard, E.; Mohamed, A.; Ross, J.; Sunarto, S. & Traeholt, C. (2015). "Prionailurus planiceps". IUCN Red List of Threatened Species. 2015: e.T18148A50662095.
doi:10.2305/IUCN.UK.2015-2.RLTS.T18148A50662095.en. Retrieved 16 January 2022.

336p - Yongcheng, L.; Richardson, M. 2021. "Rhinopithecus roxellana". IUCN Red List of Threatened Species. 2021: e.T19596A196491153.
doi:10.2305/IUCN.UK.2021-1.RLTS.T19596A196491153.en. Retrieved 14 November 2021.
- Groves, C. P. 2005. Wilson, D. E.; Reeder, D. M. eds.. Mammal Species of the World: A Taxonomic and Geographic Reference 3rd ed.. Baltimore: Johns Hopkins University Press. p. 174. ISBN 0-801-88221-4. OCLC 62265494.

338p - Cassola, F. (2016). "Callosciurus prevostii". IUCN Red List of Threatened Species. 2016: e.T3603A22253650.
doi:10.2305/IUCN.UK.2016-2.RLTS.T3603A22253650.en. Retrieved 15 November 2021.
- Lurz, P.W.W.; I. Fielding; V. Hayssen (2017). "Callosciurus prevostii (Rodentia: Sciuridae)". Mammalian Species. 49 (945): 40–50.
doi:10.1093/mspecies/sex004.

340p -아이수양, 「알고보면 서러운 비둘기의 삶 이야기」,
https://j100.co.kr/entry/알고보면-서러운-비둘기의-삶-이야기, 2023.04.17
-동아사이언스, 「[과학향기]평화의 상징 비둘기는 서럽다」,
https://www.dongascience.com/news.php?idx=-88278, 2023.04.17

342p - BirdLife International 2018. "Pithecophaga jefferyi". IUCN Red List of Threatened Species. 2018: e.T22696012A129595746.
doi:10.2305/IUCN.UK.2017-3.RLTS.T22696012A129595746.en. Retrieved November 11, 2021.
- "Appendices | CITES". cites.org. Retrieved January 14, 2022.

344p - Leary, T.; Singadan, R.; Menzies, J.; Helgen, K.; Wright, D.; Allison, A.; Aplin, K.; Dickman, C. 2016. "Spilocuscus maculatus". IUCN Red List of Threatened Species. 2016: e.T20636A21950307.
doi:10.2305/IUCN.UK.2016-2.RLTS.T20636A21950307.en. Retrieved 12 November 2021.
- Groves, C. P. 2005. Wilson, D. E.; Reeder, D. M. eds.. Mammal Species of the World: A Taxonomic and Geographic Reference 3rd ed.. Baltimore: Johns Hopkins University Press. p. 48. ISBN 0-801-88221-4. OCLC 62265494.

346p - Thorington, R. W., Jr; Hoffmann, R.S. (2005). "Sciurus (Otosciurus) aberti". In Wilson, D.E.; Reeder, D.M (eds.). Mammal Species of the World: a taxonomic and geographic reference (3rd ed.). The Johns Hopkins University Press. pp. 754–818. ISBN 0-8018-8221-4. OCLC 26158608.
- Cassola, F. (2017). "Sciurus aberti". IUCN Red List of Threatened Species. 2017: e.T42461A22245623.
doi:10.2305/IUCN.UK.2017-2.RLTS.T42461A22245623.en. Retrieved 6 September 2021.

348p - Wozencraft, W. C. 2005. "Ursus arctos arctos". In Wilson, D. E.; Reeder, D. M. eds.. Mammal Species of the World: A Taxonomic and Geographic Reference 3rd ed.. Johns Hopkins University Press. pp. 588–589. ISBN 978-0-8018-8221-0. OCLC 62265494.
- Calvignac, S.; Hughes, S 1998. "Ancient DNA evidence for the loss of a highly divergent brown bear clade during historical times" PDF. Molecular Ecology. 17 8: 1962–1970. doi:10.1111/j.1365-294x.2008.03631.x. PMID 18363668. S2CID 23361337.

350p - Kranz, A.; Abramov, A.V.; Herrero, J. & Maran, T. 2016. "Meles meles". IUCN Red List of Threatened Species. 2016: e.T29673A45203002.
doi:10.2305/IUCN.UK.2016-1.RLTS.T29673A45203002.en. Retrieved 19 November 2021.
- Kilshaw K, Newman C, Buesching CD, Bunyan J, Macdonald DW 2009. "Coordinated latrine use by European badgers, Meles meles: Potential consequences for territory defense". Journal of Mammalogy. 90 5: 1188–1198.
doi:10.1644/08-MAMM-A-200.1. JSTOR 27755113. S2CID 86435009.

352p - Groves, C. P. (2005). Wilson, D. E.; Reeder, D. M. (eds.). Mammal Species of the World: A Taxonomic and Geographic Reference (3rd ed.). Baltimore: Johns Hopkins University Press. p. 69. ISBN 0-801-88221-4. OCLC 62265494.
- Leary, T.; Seri, L.; Flannery, T.; Wright, D.; Hamilton, S.; Helgen, K.; Singadan, R.; Menzies, J.; Allison, A.; James, R. (2016). "Thylogale brunii". IUCN Red List of Threatened Species. 2016: e.T21870A21958826.
doi:10.2305/IUCN.UK.2016-2.RLTS.T21870A21958826.en. Retrieved 12 November 2021.

356~359p -새만금개발청 공식 블로그, 살아 숨쉬는 미래녹색도시 그린인프라 너가 궁금해!, https://m.blog.naver.com/smgcstory , 2023.04.14
- 윤은주,박종순,이치주,홍나은, "뉴노멀시대의 도시 그린인프라 계획모형 제안", KRIHS POLICY BRIEF No.872 (2022):
-박근현, "정원도시 싱가포르의 공원녹지체계 파크커텍터", 국토연구원, 국토 2013년 2월호 (통권376호)
--jardineriaon, 「녹색지붕이란 무엇입니까?」,
https://www.jardineriaon.com/ko/cubierta-vegetal.html, 2023.04.17
-김현지, 「코펜하겐이 옥상을 녹색으로 바꾸는 이유」,
https://www.smarttoday.co.kr/news/articleView.html?idxno=3580, 2023.04.17
-환경부, 국립생태원 생태통로네트워크,
https://www.nie-ecobank.kr/wildlifecrossing/ecocorridor/EcocorridorInfo.do, 2023.04.17
-환경부, 김혜리, 「국립공원 생태통로 이용하는 야생동물 5년간 2.5배 증가」, 2023.04.17
-http://www.me.go.kr/home/web/board/read.do?boardMasterId=1&boardId=988315&menuId=286, 2023.04.17

ENDANGERED ANIMALS GRAPHIC ARCHIVES 150

Present & Book design
Sungsil Graphics

Author / Director
Namsung Kim

Illustrator
Namsung Kim & Insil Lee

Assistant Editor
Hyemi Bae

Contact
Sungsil Graphics,1F, 85, Donggyo-ro 19-gil, Mapo-gu,
Seoul, 03999, Korea
-
www.sshwarang.net
sungsil@ssgraphics.net

5th published
2023. 4. 22

reddot award 2014
winner

ㅅㅓㅇㅅㅣㄹ GRAPHICS
DESIGN BOUTIQUE

본 출판물의 모든 글과 그림은 저작권법의 보호를 받습니다.
또한 이 출판물은 저작권자의 허가없이 다른 목적으로 사용할 수 없습니다.
잘못된 책은 구입한 곳에서 교환 가능합니다.
출판사 신고번호 제 2016-000195 호

All Rights Reserved. No part of this publication may be reproducedor
transmitted in any form or by any means, electronic or mechanical,
including photocopy, recording or any other information storage
andretrieval system, without prior permission in writing publisher.
Imperfect books are exchangeable place of sales.

ISBN 979-11-958484-2-3